Stochastic Theory of a Risk Business

Stochastic Theory of a Risk Business

HILARY L. SEAL

Yale University

John Wiley & Sons, Inc.

New York · London · Sydney · Toronto

Library of Congress Catalog Card Number: 74-76061
SBN 471 76900 2

Printed in the United States of America

TO FRANK ANSCOMBE:

Master eclectic
From sequential inspection
To Bayesian clique.

Preface

This monograph is the result of an attempt to survey all the literature relating to the mathematical foundations of risk taking as a business. And since the business cannot be run without mathematics my subject is a practical one. No other book exists that brings together the subjects included herein but I found that the relevant articles, scattered through the actuarial and statistical literature of a dozen countries, grouped themselves naturally into the topics indicated by the chapter titles and subheadings.

After describing what I understand by a risk business, I am led naturally to a description of the various probability distributions that have been proposed or utilized to describe the year-to-year payments made by such a business. My examples are mainly drawn from the insurance field: this is because information on gambling casinos or private pension and welfare funds is not readily available.

Having discussed the "outgo" in Chapter 2 I turn to the "income" in the following chapter. Each premium, as I call the fee paid by the contractholder, is divided into two portions: (a) the "net" premium which, when aggregated over all contractholders, is intended to estimate the mean of the distribution of outgo, and (b) the risk loading, familiar to all of us who have seen a roulette table. Only item (a) is dealt with in Chapter 3. A further element in each premium which consists of the proportionate part of the company's expenses of operation is ignored throughout this work. It is considered that the problem of forecasting and apportioning these expenses can be solved by techniques that are not special to a risk business.

Chapter 4 prepares the reader to decide on the size of item (b), namely the risk loading to be included in each premium. It is frankly mathematical. However those who find this theory hard going may console themselves with the thought that until the last 10 years the results could only be derived by even more difficult and circuitous methods. In this chapter more than anywhere else I draw techniques and results from outside the actuarial literature.

Ready to calculate premium loadings we discover that they are closely related to the problem of *changing* the business's probability distribution of aggregate outgo. This change may be effected by rejecting certain large individual risks or by reinsuring a portion of the company's risk portfolio. These problems are solved in Chapter 5 by considering the company's risk of ruin in the sense of Chapter 4. Alternatively, an appeal can be made to the economic theory of utility. Utility or "moral expectation" goes back to Daniel Bernoulli and without it one cannot explain why a contractholder should gladly pay a risk loading on his net premium. Once it is accepted that a risk business should have its "own" utility index (dictated, perhaps, by the consensus of attitudes towards risk taking of the company's directors) certain types and amounts of reinsurance are seen to be better than others. The theory cannot be as easily applied as that of ruin probabilities— simply because we find it so hard to ascribe numerical values to our utilities whereas we all think we know what a "small" probability of ruin is. On the other hand utility theory can demonstrate very clearly that certain procedures are illogical and thus presumably to be avoided.

Although it might have been neater to include both approaches to reinsurance in a single chapter I preferred to emphasize their antithetical character. I have thus relegated the applications of utility theory to the final chapter.

A word is necessary about the references collected together at the end of the chapters. I have tried to find and read every article written in any of the main European languages (except Russian) which has a direct bearing on my subject. The result is that I have cited over 250 articles and books in the course of this monograph. Although I believe I have commented on the principal contribution of each of the articles I do not pretend to have understood them all in the way the author intended. I have taken some pains to track down the original author of a fundamental new idea or technique, and some of my references to earlier literature may come as a shock. I have not listed about 100 of the articles I read either because they appeared to me to cover old ground or because they were of an expository or introductory character. Where cited articles have been translated into English this has been noted.

In conclusion I should mention that this publication is in no sense a textbook on the business of casualty insurance. No rules are given for deciding whether a risk is "insurable," for calculating expense margins to be included in the premiums, or for computing year-end reserves to be entered on a company's balance sheet. My concern has been with mathematical and statistical theory, not with its application in practice. On the other hand I have tried to confine myself to results that have some possibility of being applied. Purely mathematical theorems have not generally been reported.

My sincere thanks go to William DuMouchel who corrected many of my mistakes in the original classroom notes. Francis J. Anscombe, Charles C. Hewitt, Jr., and Stefan Vajda kindly read the manuscript and made numerous suggestions for its improvement. Any further comments and criticisms from readers will be gratefully received.

HILARY L. SEAL

New Haven, Connecticut
Date: March, 1969

Contents

CHAPTER 1

The Risk Business

Any business enterprise runs the risk of loss or failure, but relatively few deliberately expose themselves to losses that may exceed their receipts as an essential characteristic of the business they are engaged in. These enterprises undertake to make payments to their clients on the occurrence of certain contingencies (such as the failure to die) within a contractual period. The clients pay in advance much smaller sums called premiums or "stakes" in order to participate in the business's distributions. As early as 1738 Daniel Bernoulli indicated why it could be advantageous for a man of limited means to pay more than the mathematical expectation of his gain for such an opportunity. Even earlier DeMoivre had proved, in effect, that a risk business would eventually be ruined if it failed to include a margin in its favor in the price it charged for its contingent payments. From both viewpoints, therefore, the premiums payable to the risk business should incorporate a so-called risk-loading and may also support a proportionate part of the administrative expenses and profits of the enterprise. A business of this type is called a risk business.

Examples of risk businesses are (a) a gambling casino, (b) an employer who has guaranteed to pay his employees' medical expenses out of a trust fund set up expressly for the purpose, (c) a fire insurance company, and (d) a life insurance company that issues only short-term policies. On the other hand, we have not included among our risk businesses the investment trust fund (mutual fund, unit trust) whose sole purpose is to obtain a good return on a portfolio of investments, one reason being that the entrepreneur never ends up by owing more than he has invested. The risk situations faced by the managers of such a business are the subject of a chapter in Borch (1968). In spite of this exclusion the variety of risk businesses remains wide, and we should dearly have liked to use illustrations drawn from each of the areas in which risk businesses operate. Unfortunately almost all the mathematical theory has been developed in connection with insurance and uses its terminology. Nevertheless we have retained the expression "risk business"

1

throughout to remind readers that most of the concepts are equally applicable in all areas of risk.

Two of the above examples of risk business need amplification. In example (b) the employer's own business is not the risk business: the trust fund is the risk business and its premiums are the employer's contributions. The limitations in example (d) to short-term insurance policies was to draw the reader's attention to the fact that most life insurance companies are more than just risk businesses. By historical accident the bulk of life insurance contracts written today includes a large element of saving. In fact, each premium in a contract extending over several years may be apportioned mathematically between "risk" and "saving." Each year the "saving" element in any premium is accumulated by the insurance company, and the sum-at-risk, which is the difference between the sum insured and the accumulation, decreases. At any given epoch, then, a life insurance company has substantial accumulations of its policyholders' money which are used essentially to cover its corresponding liabilities (for cash surrender values and the like) to those policyholders. These sums have nothing to do with the "internal" risk business it is conducting. The latter, regarded perhaps as a self-supporting department of the company, is subject to the theoretical considerations we develop herein.

Now, although every premium must include a risk loading or "house cut" (as we shall later prove), it would be unwise of a risk business to rely solely on these risk loadings to pay a run of "unexpected" claims for payment. In fact, we state that the essential properties of a risk business are the following:

1. A risk reserve—a convenient expression for working capital or "free reserves."

2. A premium income and, possibly, an interest income derived from its risk reserve.

3. An outgo of claims (which we prefer to the word "losses") paid on account of the occurrence of contractual contingencies.

As an example of a legislative requirement of minimum risk reserves, we mention the case of nonlife insurance companies in Great Britain. Before 1967 any such company's minimum risk reserve was fixed at the smaller of £50,000 and 10% of gross premium income. Currently this minimum is 20% of the first £2.5 million of nonlife premium income (with a minimum of £100,000) plus 10% of the excess over £2.5 million. Another illustration of the relative size of risk reserves is provided by the 435 fire and casualty insurance companies that were doing business in New York state in 1964.*

* *The* 107th *Annual Report of the Superintendent of Insurance for Year Ended December 31, 1965*, Albany, N.Y.

The aggregate capital and surplus of these companies at the end of the year was \$14,495,643,481, premiums less expenses during 1964 amounted to \$7,621,565,075, and claims were \$8,207,994,061.

It is convenient to think of the premiums as being added to the risk reserve as they are received and to require claims to be paid from this reserve. The risk reserve at any point of time then represents the actual working capital of the enterprise at that epoch. Often, however, we consider the risk reserve only at given equidistant accounting epochs and ignore its behavior within these intervals.

For our purposes a risk business's working capital (risk reserve) must be deemed to include any money that could be borrowed by the enterprise. The reduction of this working capital to a zero, or a negative, quantity thus implies the "ruin" of the business. We use this terminology even though it strikes harshly on insurance executives' ears. Note that because the actuary generally sets a conservative (i.e., high) value on the liabilities of a life insurance company such a company with a negative risk reserve in its risk department might well be able to continue in business by revealing a surplus in its other reserves. We ignore this possibility when we speak of "ruin" in what follows.

Hereafter the term "premium" will include whatever risk loading the enterprise considers proper in the light of its profit, stability, or "utility" requirements—matters to be discussed later. However, this premium does not include any allowance for administrative or other expenses. When we wish to speak of the premium less its risk loading, we shall refer to the net or "pure" premium. We use the word "claims" to mean the actual payments made to the clients (who are sometimes the insured) rather than the possibly different amounts they claim for payment. Insurance terminology differs from branch to branch and from country to country and we cannot hope to please everybody.

REFERENCE

Borch, K. H. (1968). *The Economics of Uncertainty*. Princeton University Press, Princeton, N.J.

CHAPTER 2

The Distribution of Aggregate Claims

If we look no further than the end of the accounting period (a year, say) that is just commencing, the problem of finding a model to describe the aggregate claim outgo is familiar in statistical theory. Instead of asking for a forecast of the year's outgo, we seek the probability distribution of aggregate claims during the year. If X is the random variable of the year's claim outgo, we may write

$$\mathfrak{I}\{X \leq x\} = F(x), \qquad 0 < x < \infty, \tag{2.1}$$

and the problem becomes that of determining $F(\cdot)$. A whole spectrum of (continuous) probability distributions and standard techniques for the estimation of their parameters from actual data would seem to be available. In fact, certain difficulties present themselves. No risk business, whether an insurance company, a trust fund covering a group of employees, or a gambling casino, is a static entity. Not only is the portfolio of covered objects continually changing but the actual size of the business is subject to growth or diminution. If a long series of sample values of X is obtained from past records, a time trend will probably be manifest. There is thus a danger that if the time order of the X-values is neglected a spuriously good fit could be obtained to a series of observations.

Several alternative approaches are possible, however.

1. Relate the aggregate claim outgo X to the aggregate potential outgo (sum insured) of that year.

2. Analyze the claims into a number of categories (e.g., according to the ages of covered individuals) and relate each group of claims to the corresponding exposure.

3. Consider the number of claims as well as X and assume, possibly, that the probability distribution of the individual claim size is independent of the number of claims that occur.

4

Each of these approaches in turn leads to others. The objective, however, should not be to display ingenuity in the proliferation of alternative models but to achieve a reasonably good fit of the available data by using assumptions that can eventually be tested on further data.

CLAIM RATIO DISTRIBUTIONS

Most of the literature that provides illustrations of the fitting of series of annual claim (or loss) ratios was written with the object of finding the upper $100\,p\%$ point (pth quantile) of the resulting distribution. Such a point is needed in connection with the determination of the "proper" risk reserve and the resulting reinsurance deemed necessary.

A graphic comparison of a whole distribution of actual and expected loss ratios $R = X/S$, where S is the aggregate sum insured, was published by Riebesell (1936) and based on 626 fire insurance business periods, the expected frequencies being calculated as a Pearson Type IV curve (Kendall & Stuart, 1958); namely,

$$F'(r) \propto \left(1 + \frac{r^2}{a^2}\right)^{-m} \exp\left(-\nu \tan^{-1}\frac{r}{a}\right), \qquad -\infty < r < \infty.$$

He found that the results differed strongly from the Normal, although Sergowskij (1937), by means of four graphic comparisons, each based on a different set of 20 years' R-values, considered the Normal to be reasonably adequate.

On the other hand, Wold (1937) has referred to the successful fitting of Pearson Type III curves; namely,

$$F'(r) \propto \left(1 + \frac{r}{a}\right)^p e^{-pr/a}, \qquad -a \leq r < \infty,$$

to Swedish rural fire insurance company claim ratios. The infinite range of the foregoing distributions suggests that Rossmann's choice (1938), and that of Campagne et al. (1947), of a Pearson Type I curve was more logical. In the first of these two articles the loss ratios experienced by a French hail insurance company in the 49 years 1888–1936 were fitted by

$$F'(r) \propto (r - 0.00606)^{1.11}(0.01526 - r)^{1.42}.$$

In the second work the authors fitted each of three Dutch fire insurance companies' loss ratios for 25 years by means of Type I curves extending from zero to unity. The density function of $R = X/S$ was written as

$$\frac{\Gamma(p + 1)\,\Gamma(\alpha p + 1)}{\Gamma(\alpha + 1 \cdot p + 2)}\, r^p(1 - r)^{\alpha p}, \qquad 0 \leq r \leq 1, \tag{2.2}$$

Table 2.1

Company	$10^6 \hat{\rho}$	$10^6 \hat{\rho}_0$	$\hat{\alpha}$	$\hat{p}$
A	2132	354	2823.86	0.198251
B	234	53	188.67	0.29268
C	721	676	1478.9467	14.829129

and, instead of fitting by moments, they made use of the relations

$$\alpha = \frac{1 - \rho_0}{\rho_0} \quad \text{and} \quad p = \frac{\rho_0}{\rho - \rho_0} (1 - 2\rho_0),$$

where ρ and ρ_0 are the mean and mode, respectively, of the distribution. The estimation of ρ_0 was described in detail and the results are listed in Table 2.1. It is to be presumed that the standard errors of these estimates are relatively large. However, the very diverse results obtained—company A with an almost uniform distribution and company B with a distribution tending toward L-shape—indicate how rash it would be to hope for a single general form for $F(\cdot)$.

A more recent attempt to fit a series of annual fire insurance claim ratios for the 45 years 1916–1960 (reinsured amounts deducted from both numerator and denominator) is that of Welten (1963). The hypothesis of no trend in these ratios was tested by calculating the coefficient of correlation between r_j, the claim ratio for year j, and j itself ($j = 1, 2, \ldots, 45$). The result was -0.098, not significantly different from zero by Fisher's z-method. The member of the Pearson family of distributions that was appropriate to the calculated set of first four moments (Elderton, 1938) was Type I; namely,

$$(\text{const}) (r - a)^{m-1} (b - r)^{n-1}, \quad 0 < a < r < b, \tag{2.3}$$

where

$$10^4 \hat{a} = 0.758, \quad 10^4 \hat{b} = 6.638, \quad \hat{m} = 0.134, \quad \hat{n} = 0.392.$$

This is a U-shaped distribution and its upper tail was considered exaggerated, since there were 11 values of $10^4 r$ "expected" in excess of five, whereas only seven were observed in this range. Accordingly, a bell-shaped lognormal distribution (i.e., a distribution in which log R instead of R is Normal) was fitted which gave 4.4 "expected" values in excess of $10^4 r = 5$.* This

* The use of log $\{R/(1 - R)\}$ might appear to be a more logical choice for a variable with a range $(-\infty, \infty)$. However, F. H. Murphy, a student of mine, found that there were 4.05 "expected" values of $10^4 r$ in excess of 5 when this variable was fitted by the Normal.

illustrates how difficult it is to choose a suitable mathematical form for $F(\cdot)$, even after 45 years' results are available.

An example in which explicit account was taken of the time trend in the claim ratios is provided by Hesselberg (1937). In this case the aggregate claim ratios were based on life insurance sums-at-risk (i.e., sum insured minus reserve required) in excess of 100,000 kroner per policy covered by a mutual reinsurance pool organized by seven Norwegian life insurance companies. These claim ratios were available for each of the seven companies in respect of each of the 16 years 1920–1935. The straight line $r_i = a + bi$ $(i = 1, 2, \ldots, 16)$ was then fitted by least squares with weights equal to s_i, the aggregate sum-at-risk in year i. Reference may be made to the original paper for the estimation of σ_i^2, the variance of any r_i-value. A comparison of the 112 observed values of $r_i/\sqrt{2 \cdot \hat{\sigma}_i}$ with the corresponding expectation under the Normal distribution with zero mean and unit variance is made in Table 2.2. A chi-squared goodness-of-fit test results in a variate value 7.170 with 12 degrees of freedom. This example illustrates, perhaps, that life insurance data are more easily represented by simple models than the corresponding fire and casualty statistics. In this connection we should remember that the amount of a fire insurance claim, for example, is itself a random variable whose size is determined only after the fire has occurred.

Table 2.2

Value of $r_i/\sqrt{2 \cdot \hat{\sigma}_i}$	Observed Frequency	Normal Frequency
$-\infty$ to -2.00	—	0.26
-2.00 to -1.75	—	0.48
-1.75 to -1.50	2	1.15
-1.50 to -1.25	—	2.42
-1.25 to -1.00	5	4.49
-1.00 to -0.75	9	7.37
-0.75 to -0.50	10	10.68
-0.50 to -0.25	18	13.67
-0.25 to 0.00	12	15.48
0.00 to 0.25	18	15.48
0.25 to 0.50	15	13.67
0.50 to 0.75	8	10.68
0.75 to 1.00	7	7.37
1.00 to 1.25	5	4.49
1.25 to 1.50	1	2.42
1.50 to 1.75	1	1.15
1.75 to 2.00	1	0.48
2.00 to ∞	—	0.26
	112	112.00

As a further example we mention the 100 fire insurance claim ratios (1857–1956) and the 69 hail insurance claim ratios (1888–1956) fitted log-normally by Ramel (1960). In the first series the trend was slowly downward and in the second (in which the fluctuations were much wider) the trend was upward. Unfortunately the denominators of these ratios were the insurance companies' own (gross) premiums and, since they were presumably changed from time to time based on experience, the successive terms are likely to be negatively correlated with too small a variability.

Finally we refer to the use in workmen's compensation insurance of the "excess pure premium ratio"

$$\phi(r) = \kappa_1^{-1} \int_{r\kappa_1}^{\infty} (x - r\kappa_1) \, dF(x), \qquad (2.4)$$

where $\kappa_1 = \int_0^\infty x dF(x)$. Here the various groups of covered work men in a given risk category ("premium class") produce the observed values of X during a given year and can be divided by a mean loss figure to obtain the corresponding values of $X/\hat{\kappa}_1$. By grouping these ratios into a histogram it is then possible to derive an observed set of ordinates $\phi(r)$ for suitably chosen r-values ($0 < r < \infty$). The problem is then to produce a smooth or graduated set of $\phi(r)$ values by mathematical formula or otherwise. Since $\phi'(r) = F(r) - 1$, the graduation of $\phi(r)$ is equivalent to graduating $F(x)$. Two examples may be mentioned: (a) Valerius (1942), who graduated 14 separate premium-size classes by the Whittaker-Henderson method (Camp, 1950), and (b) Simon (1965), who reported the graduation procedure used for 36 standard premium classes encompassing an aggregate of 112,646 groups of workmen: for most of the premium classes Simon fitted an eighth-degree polynomial to $[\phi(r)]^{-1}$ but in some cases he fitted Type III curves (with nonzero origin) to $F(r)$. A more sophisticated approach is that of Hewitt (1967), who proposed a gamma distribution with unit mean for $\phi(r)$ and illustrated its successful fitting to 1958 California data with premiums between \$25,000 and \$50,000.

Reviewing the foregoing attempts to provide models for R, the random variable of loss ratios, we are led to conclude that the very diversity of results is an argument against such a "simple" approach.

DISTRIBUTIONS BASED ON WITHIN-PORTFOLIO CLASSIFICATION

A feature of the foregoing specification of a d.f. for R was that, being based on a number of past years' experiences, it was theoretically capable of being extended over several (even many) years into the future. It thus gave management an opportunity to predict the results of future years' operations, provided estimates were available of the likely (or possible) future development of the aggregate portfolio S of sums insured.

We now turn to a model that can supply only an estimate of $F(\cdot)$ for the accounting period about to commence. It is based on a classification of the individual risks of a portfolio by grouping together all risks with the same value of a given meaningful criterion. This type of classification has been much utilized in life insurance in which the overriding consideration is the age of the life insured. Suppose, for example, that n coeval lives are insured for a unit of money in case of death within the year. If q is the (unknown) probability of dying for each of these lives independently of one another, we have the binomial distribution of deaths (Aitken, 1957); namely,

$$\mathfrak{I}\{X \leq x\} = F(x) = \sum_{j=0}^{[x]} \binom{n}{j} q^j (1 - q)^{n-j}. \tag{2.5}$$

The problem of specifying $F(\cdot)$ has, for this group, been reduced to that of estimating q from past experience. In effect, the "whole portfolio" of observations used in the R-approach is here replaced by observations relating solely to claim experience in any given classification. It is well known that as n increases without limit

$$F(x) \sim \Phi\left(\frac{[x] + \frac{1}{2} - nq}{\sqrt{nq(1 - q)}}\right), \tag{2.6}$$

where

$$\Phi(z) = \frac{1}{\sqrt{2\pi}} \int_{-\infty}^{z} e^{-t^2/2}\, dt.$$

The addition of the $\frac{1}{2}$ in the numerator of (2.6) was first suggested by De Morgan (1837).

If an insurance company were, in fact, limited to lives of a single age (or a narrow range of ages) and coverage remained relatively steady at N lives—a conceivable situation for an employer of unskilled labor—the foregoing model would apply over a longer period than a single year. A more practical model, however, would be to assume that the company is composed of k homogeneous groups of lives, there being n_j lives in group j, all subject to a mortality rate $q_j (j = 1, 2, \ldots, k)$. If the deaths within each group occur independently of one another and independently of the deaths in each of the other groups, the distribution function of the aggregate deaths is the binomial of Poisson (Aitken, 1957), namely

$$F(x) = \text{sum of the coefficients of } \theta^0, \theta^1, \theta^2, \ldots, \theta^{[x]} \text{ in } \prod_{j=1}^{k} [q_j + (1 - q_j)\theta]^{n_j} \tag{2.7}$$

$$\sim \Phi\left(\frac{[x] + \frac{1}{2} - n\bar{q}}{\sqrt{n\bar{q}(1 - \bar{q}) - n\sigma_q^2}}\right), \tag{2.8}$$

where

$$n = \sum_{j=1}^{k} n_j, \qquad \bar{q} = \sum_{j=1}^{k} \frac{n_j q_j}{n}, \quad \text{and} \quad \sigma_q^2 = \sum_{j=1}^{k} \frac{n_j (q_j - \bar{q})^2}{n}.$$

Note that the "correction" $n\sigma_q^2$ to the assumption that all lives have a probability of death within the year equal to $\bar{q}$ is likely to be relatively small; for example, if $n_1 = n_2 = n_3 = 100$ and $q_1 = 0.01$, $q_2 = 0.02$, $q_3 = 0.03$,

$$n\bar{q}(1 - \bar{q}) = 5.88 \quad \text{and} \quad n\bar{q}(1 - \bar{q}) - n\sigma_q^2 = 5.86$$

(Walsh, 1955).

To remove the limitation we have imposed on the sums insured (i.e., they are all equal to unity) we have to consider the k triplets (n_j, s_j, q_j), where s_j is now the sum insured (or the sum-at-risk) of each of the n_j contracts in group j. If these k triplets are regarded as fixed parameters, the random variable X is the sum of a linear combination of k independent random variables X_j, where $(j = 1, 2, \ldots, k)$

$$\mathcal{P}\{X_j = x_j\} = \binom{n_j}{x_j} q_j^{x_j} (1 - q_j)^{n_j - x_j}, \qquad x_j = 0, 1, 2, \ldots, n_j.$$

In fact,

$$X = \sum_{j=1}^{k} s_j X_j,$$

so that

$$\mathcal{E}X = \sum_{j=1}^{k} s_j \mathcal{E} X_j = \sum_{j=1}^{k} s_j n_j q_j = \mu, \text{ say,}$$

$$\mathcal{U}X = \sum_{j=1}^{k} s_j^2 \mathcal{U} X_j = \sum_{j=1}^{k} s_j^2 n_j q_j (1 - q_j) = \sigma^2, \text{ say,}$$

and, in general, if $\kappa_l(X)$ is the lth cumulant of the distribution of X and $\kappa_l(n_j, q_j)$ is the lth cumulant of the binomial distribution with parameters n_j and q_j (Kendall & Stuart, 1958),

$$\kappa_l(X) = \sum_{j=1}^{k} s_j^l \kappa_l(n_j, q_j). \tag{2.9}$$

It is thus possible to fit various types of frequency distribution to aggregate X by the method of moments. As $n \to \infty$, the Central Limit theorem states that

$$F(x) \sim \Phi\left(\frac{x - \mu}{\sigma}\right),$$

and this (Normal) distribution has been used by Pizzetti (1964) on 1000 one-year term insurances with nine different "central" sums insured. Taylor (1967), on the other hand, using a standard set of values of q_j, fitted Pearson curve types III and IV as well as the Normal to two distributions of sums-at-risk on 19,752 and 83,057 lives, respectively. As might be expected, the discrepancies between the three fitted distributions were more striking at the tails. We may note that only a single "observation" was possible for each of these two groups of lives, namely, the actual value of X for the year immediately following the calculations. At the end of that year most n_j and every s_j and q_j will have assumed new values.

The foregoing model can, of course, be applied to the whole duration of life insurance contracts that extend many years into the future and are subject to the payment of stipulated annual or other premiums. In such a case, for example, L_t could represent the random variable of the present value of the amount of loss (this loss being the sum insured payable minus the accumulated net premiums) in case of death during the tth year hence, the probability of death during that year being a function of t and x, the life insured's present age. The probability distribution of the (present value of the) aggregate amount paid through time t is then the product of a series of multinomial distributions, one for each (independent) contract (Stone, 1948). Great ingenuity was displayed by European continental writers of the nineteenth and early twentieth centuries in devising moment formulas (usually only up to the second order) for more and more general types of life insurance contract. Readers interested in algebraic manipulations are referred to Bohlmann (1909) for the earlier history, Berger (1939) for the later generalizations, Hickman (1964) for the case of multiple decrements (e.g., disability as well as death), Williams (1948) for an elementary approach to moments of higher order, and Seal (1953) for a practical application of variance minimization in the design of pension plans.

An example of fitting the first three aggregate moments of the monthly annuity payments made until the deaths of a closed group of 22 lives is given by Taylor (1952). The theoretical distribution chosen was a Pearson Type III curve.

As we have emphasized, the "individual contract" model adumbrated above permits only a very short range forecast of the behavior of X. In particular, if it were to be applied to casualty rather than life insurance, the short-term nature of the contracts and the rapid portfolio changes would essentially limit application to one year. Whereas the R-model was stigmatized as inaccurate, we could perhaps say that the present model is too precise. A way of removing the necessity of having to specify the set $\{s_j\}$ for every future year over which a forecast is desired would be to consider the sum insured (or sum-at-risk) as a random variable. Any contract with risk probability q_j would then be given a sum insured selected at random from the distribution of sums insured—which could depend on j. It is then possible to derive the probability distribution of aggregate claims in which each claimant has a sum insured that is a variate from the (discrete) probability distribution of sums insured (Seal, 1947, 1951, where sums insured are replaced by numbers of policies per life insured).

INDIVIDUAL CLAIM AMOUNTS A RANDOM VARIABLE

Let us pursue the idea that individual sums insured are subject to a probability distribution. The concept that a claim on an insurance company consists of two independent events, the occurrence of the claim and its amount, goes back to nineteenth-century attempts to provide theories of sickness indemnity and fire insurance, but little progress was made until it was discovered that the Poisson distribution was a satisfactory way of representing the probability of n claims occurring in a given portfolio of contracts within a specified period of length t. Write this probability as $p_n(t)$. If Y, the size of an individual claim, is a positive random variable that is independent of the random variable number-of-claims and its distribution function $P(y)$ does not involve t, then

$$\mathcal{F}\{X \leq x\} \equiv F(x, t) = \sum_{n=0}^{\infty} p_n(t)\, P^{n*}(x), \qquad (2.10)$$

where $P^{n*}(x)$, the nth convolution of $P(x)$, is given by

$$P^{n*}(x) = \int_0^x P^{(n-1)*}(x - y)\, dP(y), \qquad n = 1, 2, 3, \ldots, \qquad (2.11)$$

is the distribution function of the sum of n individual claims and is obtained recursively by noting that $dP(y)$ is the probability of a claim of size between y and $y + dy$. To complete the definition we require $P^{1*}(x) = P(x)$, $P(0) = 0$, and, since this implies that $P^{n*}(0) = 0$ while $F(0, t) = p_0(t)$, we define

$$P^{0*}(x) = \begin{cases} 0, & x < 0, \\ 1, & x \geq 0. \end{cases}$$

[Although we have used the notation of the Stieltjes integral (e.g., see Widder, 1961) because it will be useful when we consider the special case in which all sums insured are equal, the reader may confine himself to the continuous case in which

$$dP(y) = \frac{dP(y)}{dy}\, dy$$

and use summations when a discrete variable is considered.]

We note that if Y depends on t and has d.f. $P(y, t)$ then relation (2.10) still applies, provided

$$P(y) = \frac{1}{t} \int_0^t P(y, \tau)\, d\tau.$$

As an example, we may suppose that the aggregate claim X by time t includes interest at a force δ so that the probability of an individual claim which occurred at time τ being less than or equal to y at time t is $P(ye^{-\delta\overline{t-\tau}})$, and to apply (2.10) we must replace $P(y)$ with

$$\frac{1}{t}\int_0^t P(ye^{-\delta\overline{t-\tau}})\,d\tau.$$

Suppose X is composed of n individual independent claims such that $X^{(n)} = \sum_{j=1}^n Y_j$. Then $X^{(n)}$ has d.f. $P^{n*}(x)$. If we write the moment generating function (mgf) of Y as (Kendall & Stuart, 1958)

$$\mathcal{E}e^{\theta y} = M_Y(\theta) = \int_0^\infty e^{\theta y}\,dP(y),$$

$$\mathcal{E}e^{\theta X^{(n)}} = [M_Y(\theta)]^n = \int_0^\infty e^{\theta x}\,dP^{n*}(x),$$

and, when X is not limited to a fixed number of claims,

$$M_X(\theta) = \mathcal{E}e^{\theta X} = \int_0^\infty e^{\theta x}\,d_x\,F(x,t) = \sum_{n=0}^\infty p_n(t)\int_0^\infty e^{\theta x}\,dP^{n*}(x)$$

$$= \sum_{n=0}^\infty p_n(t)[M_Y(\theta)]^n. \tag{2.12}$$

This result can also be proved directly from the definition of $P^{n*}(x)$ in (2.11).

In order to utilize (2.10) or (2.12) it is necessary to give explicit forms to $p_n(t)$ and $P(y)$. A review of the wide range of theoretical possibilities available is given by Kupper (1960, 1962).

The Discrete Distribution $p_n(t)$

If the occurrence of a claim leaves the insured immediately liable for another (e.g., as in insurance against theft of personal belongings), it is reasonable to assume that the probability of a claim at time τ is $\lambda\,d\tau + o(d\tau)$ so that only one claim can occur at any given instant. If this model were to be applied to a portfolio in which a claim resulted in the removal of the insured (e.g., as in life insurance), it would have to be assumed that any contract terminating as a claim would be immediately replaced by an identical contract. This may not be farfetched with large portfolios subject to many accessions and severances.

Write
$$\mathcal{P}\{n \text{ claims before epoch } \tau\} \equiv p_n(\tau);$$

then, essentially following the arguments of Lundberg (1903),

$$p_n(\tau + d\tau) = p_n(\tau)\{1 - \lambda\,d\tau\} + p_{n-1}(\tau)\lambda\,d\tau + o(d\tau), \qquad n = 1, 2, 3, \dots,$$

since n claims before epoch $\tau + d\tau$ can arise only from the mutually exclusive events (a) n claims before τ and no claim in the interval $(\tau, \tau + d\tau)$, or (b) less than n claims before τ and the complement in the interval $(\tau, \tau + d\tau)$. Because the probability of two or more claims in an interval of length $d\tau$ is $o(d\tau)$, the second term is concerned only with $(n - 1)$ claims before epoch τ.

The foregoing equation may be written, as $d\tau \to 0$,

$$\lim_{d\tau \to 0} \frac{p_n(\tau + d\tau) - p_n(\tau)}{d\tau} = p_n'(\tau) = \lambda p_{n-1}(\tau) - \lambda p_n(\tau), \qquad n = 1, 2, 3, \ldots ,$$

(2.13)

and the corresponding equation for $n = 0$ is seen to be

$$p_0'(\tau) = -\lambda p_0(\tau);$$

that is,

$$\frac{d \ln p_0(\tau)}{d\tau} = -\lambda$$

or

$$p_0(\tau) = e^{-\lambda\tau}.$$

(2.14)

It is easily verified that the solution of (2.13) is

$$p_n(\tau) = e^{-\lambda\tau} \frac{(\lambda\tau)^n}{n!}, \qquad n = 0, 1, 2, \ldots .$$

(2.15)

This is a Poisson distribution with mean $\lambda\tau$, namely, the expected number of claims in the interval $(0, \tau)$. As is well known, the variance of this distribution (and, in fact, the cumulant of any order) is equal to the mean.

Two alternative models may be mentioned (Seal, 1949). If λ is a function of τ instead of being constant, the Poisson law (2.15) is still applicable but assumes the form

$$p_n(\tau) = \exp\left[-\int_0^\tau \lambda(\sigma)\, d\sigma\right] \frac{\left[\int_0^\tau \lambda(\sigma)\, d\sigma\right]^n}{n!}, \qquad n = 0, 1, 2, \ldots . \quad (2.16)$$

On the other hand, if depletions in a portfolio of N contracts due to claims (e.g., deaths) during the year are explicitly recognized, the first term on the right of (2.13), for example, becomes $(N - n + 1)\lambda\, p_{n-1}(\tau)$ and $p_n(t)$ is found to be a binomial with $q = e^{-\lambda t}$.

It will be noticed that by making a time transformation such that

$$t = \int_0^\tau \lambda(\sigma)\, d\sigma$$

relation (2.16) becomes

$$p_n(t) = e^{-t} \frac{t^n}{n!}, \qquad n = 0, 1, 2, \ldots ,$$

and depends on the sole parameter t, the expected number of claims during the interval under consideration. This transformation, which first appeared in Lundberg (1903), is a powerful tool in the forecast of the distribution of X, since it replaces a detailed knowledge of the constitution and prospects of the portfolio of risk contracts by an estimate of the "expected" number of claims during the future period considered. In fact, we may replace probability statements relating to a specified future interval of time (e.g., 5 or 10 years) with a statement about the distribution of aggregate claims in "the period during which the expected number of claims will amount to t." If t is large, the relative error made by deleting the word "expected" in the foregoing phrase will be small.

A disadvantage of the Poisson assumption for $p_n(t)$ is that the variance of the number of claims in $(0, t)$ is equal to the expected number of claims. In practice the variance is frequently larger than the mean and to meet this requirement a two-parameter distribution has been suggested (Ammeter, 1948). This is the negative binomial, namely,

$$p_n(t) = \binom{\alpha + n - 1}{n}(1 - q)^\alpha q^n,$$

$$n = 0, 1, 2, \ldots; \qquad \alpha > 0, \qquad q = \left(1 + \frac{t}{\alpha}\right)^{-1}, \quad (2.17)$$

which has mean and variance equal to $\alpha q/(1 - q)$ and $\alpha q/(1 - q)^2$, respectively. When $\alpha \to \infty$ with αq fixed, this is the Poisson distribution with mean αq.

The choice of this particular two-parameter generalization of the Poisson distribution is based on its convenience.* However, it occurs quite naturally when we consider, not the probability of n claims in a portfolio of risks but the probability of a single individual having n claims during a period of time t; for example, the distribution has been fitted satisfactorily to the numbers of automobile accidents occurring to individual policyholders by Dropkin (1959), Delaporte (1960),† Simon (1961), and Derron (1962), where the Poisson was often conspicuously unsuccessful.

As an example, the observations in Table 2.3 are the accident claims suffered in a single year by *La Royale Belge* on 9461 contracts covering two

---

* One of its properties is that if n is actually the sum of k independent random variables each distributed as the negative binomial (2.17) then n itself, which now refers to a period of length kt, has a negative binomial distribution with parameters q and $k\alpha$. Efforts have been made to use this property to apply (2.17) to a risk business with aggregate hazard varying randomly and independently for each successive interval t.

† Although this author provided the theory for the case in which $p_n(\cdot)$ is the convolution of a negative binomial and a Poisson variable (Kupper, 1960), he seems to have used only the negative binomial in his numerical illustrations.

Table 2.3

Number of Claims Made n	Number of Contracts f_n	Fitted Distributions		
		Poisson	Negative Binomial	Thyrion (1961)
0	7840	7635.62	7846.93	7840.0
1	1317	1636.72	1288.48	1322.1
2	239	175.42	256.52	225.4
3	42	12.54	54.05	51.2
4	14	0.67	11.70	14.6
5	4	0.03	2.58	4.8
6	4		0.57	1.7
7	1	—	0.17	0.7
	9461	9461.00	9461.00	9460.5

classes of "business and tourism" automobiles (Thyrion, 1961). Unbiased estimates of the mean and variance of the distribution are 0.21435366 and 0.28893137. Using the mean as the estimate of the Poisson parameter we obtain the expected frequencies shown. The fit is very bad.

To fit the negative binomial we used the maximum likelihood estimates of α and q, namely (Anscombe, 1950),

$$\frac{\hat{\alpha}\hat{q}}{1 - \hat{q}} = \text{the mean number of claims} = \bar{n} = 0.21435366$$

and $\hat{\alpha}$ is given as the solution of the equation

$$\sum_{n=1}^{\infty} f_n[\psi(\hat{\alpha} + n) - \psi(\hat{\alpha})] = \left(\sum_{n=0}^{\infty} f_n\right) \ln\left(1 + \frac{\bar{n}}{\hat{\alpha}}\right),$$

where

$$\psi(z) = \frac{d[\ln \Gamma(z)]}{dz} = \psi(z + 1) - \frac{1}{z}$$

is the psi (or digamma) function and has been extensively tabulated in Davis (1963). The initial value used for $\hat{\alpha}$ in the solution of the equation was

$$\frac{\bar{n}^2}{(\text{variance of } n) - n} \simeq 0.6,$$

and iteration with the successive values 0.7, 0.71, 0.70155 and 0.701824 produced

$$\hat{\alpha} = 0.701823$$

and thus

$$\frac{\hat{q}}{1 - \hat{q}} = 0.3054241$$

or

$$\hat{q} = 0.2339654.$$

The resulting fit is judged poor, since $\chi_3^2 = 14.69$.

Finally, we turn to Thyrion's fitting of the three parameter (mixed Poisson) distribution

$$p_n(t) = (-1)^n \frac{t^n}{n!} p_0^{(n)}(t),$$

where

$$p_0(t) = \exp \left\{ \frac{p}{c(1 - a)} [1 - (1 + ct)^{1-a}] \right\}, \qquad p, a, c > 0; \qquad a \neq 1.$$

The parameter p was estimated by $\bar{n}$ and $(ac + 1)b$ by the variance; finally, a was determined by equating $p_0(1)$ to $f_0/\sum_{n=0}^{\infty} f_n$. The numerical results taken from Thyrion (1961) produce $\chi_4^2 = 6.26$, which is an adequate fit.

An awkward characteristic of the negative binomial applied to individual contracts is that it can arise as a result of very different chance mechanisms. What is worse is that these mechanisms could lead to rather different strategies to control claims by the managers of a risk business.

Suppose first the the group exposed to risk of accident is heterogeneous in such a way that the probability that any individual chosen at random will have n accident claims within a year is given by

$$e^{-\lambda t} \frac{(\lambda t)^n}{n!}, \qquad n = 0, 1, 2, \ldots,$$

where λ, the accident hazard, is the value assumed by a random variable Λ which has d.f. $U(\cdot)$. Then the probability that a person chosen at random will have n claims in a year is the mixed Poisson

$$p_n(t) = \int_0^{\infty} e^{-\lambda t} \frac{(\lambda t)^n}{n!} \, dU(\lambda), \tag{2.18}$$

where $U(\cdot)$ may be referred to as the mixing distribution (Haight, 1967). Let us write $G(z)$ for the probability generating function (pgf) of $U(\cdot)$; that is,

$$G(z) = \int_0^{\infty} z^{\lambda} \, dU(\lambda).$$

Further, we define the factorial moment generating function (fmgf) of a discrete random variable K with probability distribution $p_k(k = 0, 1, 2, \ldots)$

as (Aitken, 1957)

$$A(s) = \sum_{k=0}^{\infty}(1 + s)^k p_k = \sum_{k=0}^{\infty} p_k \sum_{j=0}^{k}\binom{k}{j}s^j = \sum_{j=0}^{\infty}\frac{s^j}{j!}\sum_{k=j}^{\infty}k^{(j)}p_k,$$

where $k^{(j)} = k(k - 1) \cdots (k - j + 1)$ and $\sum_{k=j}^{\infty} k^{(j)} p_k$ is the jth factorial moment of the random variable K.

We notice that when p_k is the Poisson with parameter λt

$$A_P(s) = e^{-\lambda t} \sum_{k=0}^{\infty}(1 + s)^k \frac{(\lambda t)^k}{k!} = e^{-\lambda t+\lambda t(1+s)} = e^{\lambda ts}. \qquad (2.19)$$

Further, the fmgf of the mixed Poisson (2.18) is

$$A_{MP}(s) = \sum_{k=0}^{\infty}(1 + s)^k \int_0^{\infty} e^{-\lambda t}\frac{(\lambda t)^k}{k!}\,dU(\lambda) = \int_0^{\infty} e^{\lambda ts}\,dU(\lambda)$$

$$= M_U(ts), \qquad \text{the mgf of } U. \qquad (2.20)$$

We now find the mixed Poisson (2.18) in the particular case in which $U(\cdot)$ is a gamma distribution with index m and scale c^{-1}; that is,

$$dU(\lambda) = \frac{c^m}{\Gamma(m)}e^{-c\lambda}\lambda^{m-1}\,d\lambda, \qquad 0 < \lambda < \infty.$$

Then

$$M_U(ts) = \frac{c^m}{\Gamma(m)}\int_0^{\infty}e^{-(c-ts)\lambda}\lambda^{m-1}\,d\lambda = \frac{c^m}{(c - ts)^m} = \left(1 - \frac{ts}{c}\right)^{-m}$$

$$= \left(\frac{t + c}{c} - t\frac{1 + s}{c}\right)^{-m} = \left(\frac{t + c}{c}\right)^{-m}\left(1 - t\frac{1 + s}{t + c}\right)^{-m}$$

$$= \left(\frac{c}{t + c}\right)^m \sum_{k=0}^{\infty}\binom{m + k - 1}{k}\left(\frac{t}{1 + c}\right)^k(1 + s)^k,$$

and since, by (2.20), this is the fmgf of the mixed Poisson

$$p_n(t) = \binom{m + n - 1}{n}\left(1 - \frac{t}{t + c}\right)^m\left(\frac{t}{t + c}\right)^n,$$

namely, (2.17), with $\alpha = m$ and $q = t/(t + c)$ (Greenwood & Yule, 1920).

As a different model, suppose that the probability of a claim occurring between time τ and $\tau + d\tau$ is $\lambda_n(\tau)$, a function of n, the number of claims that have already occurred. The relation (2.13) now becomes (McKendrick, 1926)

$$\frac{\partial p_n(\tau)}{\partial \tau} = \lambda_{n-1}(\tau) - p_{n-1}(\tau) - \lambda_n(\tau)\,p_n(\tau), \qquad n = 1, 2, 3, \ldots, \qquad (2.21)$$

whereas, when $n = 0$,

$$p_0'(\tau) = -\lambda_0(\tau)\, p_0(\tau)$$

so that

$$p_0(t) = \exp\left(-\int_0^t \lambda_0(\tau)\, d\tau\right).$$

We consider the particular case in which $\lambda_n(\tau)$, the force of claim occurrence, is a linear function of n, say $\lambda_n(\tau) = a + bn$.* By use of the pgf defined by

$$G(z, t) = \sum_{n=0}^{\infty} z^n\, p_n(t)$$

(2.21) provides, using a standard notation for partial derivatives,

$$\frac{\partial G(z, t)}{\partial t} \equiv G_t(z, t) = -a\, p_0(t) + \sum_{n=1}^{\infty} z^n[(a + \overline{bn - 1})p_{n-1}(t) - (a + bn)\, p_n(t)]$$

$$= -a\, p_0(t) + az\, G(z, t) + bz^2 G_z(z, t) - a[G(z, t) - p_0(t)] - bz\, G_z(z,t),$$

or

$$G_t(z, t) - bz(z - 1)\, G_z(z, t) - a(z - 1)\, G(z, t) = 0. \qquad (2.22)$$

This linear partial differential equation can be solved by standard methods, but as a convenient introduction to a technique that is utilized in Chapter 4 we solve it by a transformation analogous to the mgf and known as the Laplace transform (Widder, 1961). The Laplace transform of a function $g(\cdot)$ defined on the positive real axis and tending to zero at infinity at least as quickly as e^{-sx} $[\Re(s) > 0]$ is defined by

$$\mathcal{L}[g(x)] \equiv \gamma(s) = \int_0^{\infty} e^{-sx}\, g(x)\, dx.$$

It can be proved that to any given $\gamma(s)$ there corresponds a unique $g(\cdot)$, and we note in passing that this means that $p_0(t)$ uniquely defines the mixing distribution $U(\cdot)$ in (2.18). We now write

$$\gamma(z, s) = \int_0^{\infty} e^{-st}\, G(z, t)\, dt.$$

Then

$$\int_0^{\infty} e^{-st}\, G_t(z, t)\, dt = e^{-st}\, G(z, t)\Big|_0^{\infty} + s\int_0^{\infty} e^{-st}\, G(z, t)\, dt$$

$$= -G(z, 0) + s\, \gamma(z, s)$$

$$= -1 + s\, \gamma(z, s), \quad \text{since } p_n(0) = \begin{cases} 1, & n = 0, \\ 0, & n \neq 0. \end{cases}$$

* The apparently more general assumption $\lambda_n(\tau) = (a + bn)/(A + B\tau)$ reduces to this by a change in the time scale such that $(1/B)\ln(1 + B\tau/A) = t$ (Lundberg, 1940).

On multiplying (2.22) by e^{-st} and integrating term-by-term between zero and infinity we get

$$[-1 + s\,\gamma(z, s)] - bz(z - 1)\frac{\partial}{\partial z}\gamma(z, s) - a(z - 1)\gamma(z, s) = 0,$$

that is,

$$bz(1 - z)\frac{\partial}{\partial z}\gamma(z, s) + [a(1 - z) + s]\gamma(z, s) - 1 = 0. \qquad (2.23)$$

This equation may be solved by writing

$$\gamma(z, s) = \sum_{j=0}^{\infty} c_j z^j$$

and substituting into (2.23) so that

$$bz(1 - z)\sum_{j=1}^{\infty} jc_j z^{j-1} + (\overline{a + s} - az)\sum_{j=0}^{\infty} c_j z^j - 1 = 0.$$

On comparing the coefficients of $z^0, z, z^2, \ldots$, we get

$$(a + s)c_0 = 1,$$
$$bc_1 + (a + s)c_1 - ac_0 = 0,$$

and $(j = 2, 3, 4, \ldots)$

$$bjc_j - b(j - 1)c_{j-1} + (a + s)c_j - ac_{j-1} = 0.$$

Hence

$$c_0 = (a + s)^{-1}, \qquad c_1 = a(a + s)^{-1}\,(b + a + s)^{-1}$$

and, generally,

$$c_j = \frac{b\overline{j - 1} + a}{bj + a + s}c_{j-1} = \frac{(b\overline{j - 1} + a)(b\overline{j - 2} + a)}{(bj + a + s)(b\overline{j - 1} + a + s)}c_{j-2} = \cdots$$

$$= \frac{\prod_{k=1}^{j-1}(b\overline{j - k} + a)}{\prod_{k=0}^{j-2}(b\overline{j - k} + a + s)}c_j = \frac{\prod_{k=1}^{j}(b\overline{j - k} + a)}{\prod_{k=0}^{j}(b\overline{j - k} + a + s)},$$

$$j = 1, 2, 3, \ldots.$$

Now on reference to a table of Laplace transforms (e.g., Roberts & Kaufman, 1966) or by trial and error

$$\int_0^{\infty} e^{-st}\left[\frac{1}{Q(0)} + \sum_{m=1}^{j}\frac{e^{b_m t}}{b_m\,Q_m(b_m)}\right]dt = \frac{1}{s\,Q(s)},$$

where

$$Q(s) = \prod_{m=1}^{j}(s - b_m) \quad \text{and} \quad Q_m(s) = \frac{Q(s)}{s - b_m}.$$

In this case, by replacing $s + a$ with s',

$$\sum_{m=1}^{j} \frac{e^{b_m t}}{b_m\, Q_m(b_m)} = \sum_{m=1}^{j} \frac{e^{-mbt}}{b^j (-1)^m m! (j-m)!} = \frac{1}{b^j j!} \sum_{m=1}^{j} \binom{j}{m}(-e^{-bt})^m$$

and

$$\frac{1}{Q(0)} = \frac{1}{b^j j!}.$$

Hence $(j = 1, 2, 3, \ldots)$

$$\int_0^\infty e^{-s't} \frac{(1 - e^{-bt})^j}{b^j j!}\, dt = \left[\frac{1}{\displaystyle\prod_{k=0}^{j} (b\overline{j} - k + s')} \right]^{-1}$$

and

$$\gamma(z, s) \equiv \int_0^\infty e^{-st} G(z, t)\, dt \equiv \sum_{j=0}^\infty c_j z^j$$

$$= \int_0^\infty e^{-st} \left[e^{-at} + \sum_{j=1}^\infty z^j e^{-at} \frac{(1 - e^{-bt}) \displaystyle\prod_{k=1}^{j} (b\overline{j} - k + a)}{b^j j!} \right]$$

$$= \int_0^\infty e^{-st} \left[e^{-at} + \sum_{j=1}^\infty z^j e^{-at} \binom{-a/b}{j} (e^{-bt} - 1)^j \right] dt.$$

The expression in brackets is $G(z, t)$ and the coefficient of z^j in it is $p_j(t)$. Hence

$$p_n(t) = e^{-at} \binom{\dfrac{a}{b} + n - 1}{n} (1 - e^{-bt})^n, \qquad n = 0, 1, 2, \ldots.$$

Here again we have found (2.17), but this time with $\alpha = a/b$ and $q = 1 - e^{-bt}$. Note that b must be positive, which implies that $\lambda_n(\tau)$ is an increasing function.

The second of these two models has introduced the concept of a positive correlation between successive accidents, since the occurrence of an accident increases the probability of another. It was first suggested by Pólya for "contagious" events and applied successfully by his student Eggenberger (1924) to the numbers of deaths from boiler explosions in Prussia and from smallpox and scarlet fever in Switzerland. The first of these three distributions would seem to be more aptly produced by a third chance mechanism in which accidents occur independently of one another and in which the number of accidents that result in j deaths is Poisson distributed with parameter λ_j. Such a model was proposed by Pollaczek-Geiringer (1928), who showed that if n now represents the aggregate number of *deaths* occurring in a year under an accident contract

$$p_n(1) = \exp\left(\sum_{j=1}^{\infty} \lambda_j\right) \sum \frac{\prod_{j=1}^{\infty} \lambda_j{}^{n_j}}{\prod_{j=1}^{\infty} n_j!},$$

the summation being extended over all sets $\{n_j\}$ such that $\sum_{j=1}^{\infty} jn_j = n$. It may easily be shown (Lüders, 1934) that this probability corresponds to the probability generating function

$$G_L(z) = \exp\left[\sum_{j=1}^{\infty} \lambda_j(z^j - 1)\right].$$

If the series of parameters λ_j can be represented by a logarithmic series (Lüders, 1934) so that

$$\lambda_j = \frac{hd^{j-1}}{j(1+d)^j}, \qquad j = 1, 2, 3, \ldots,$$

then

$$G_L(z) = \exp\left[\frac{h}{d}\left\{\ln\left(1 - \frac{d}{1+d}\right) - \ln\left(1 - \frac{zd}{1+d}\right)\right\}\right]$$

$$= \left(1 - \frac{d}{1+d}\right)^{h/d}\left(1 - \frac{zd}{1+d}\right)^{-h/d}$$

and

$$p_n(1) = \binom{\frac{h}{d} + n - 1}{n}\left(1 - \frac{d}{1+d}\right)^{h/d}\left(\frac{d}{1+d}\right)^n, \qquad n = 0, 1, 2, \ldots,$$

which is once again the negative binomial (2.17) but now with $\alpha = b/d$ and $q = d/(1 + d)$. Lüders fitted the negative binomial to deaths from road-rail accidents and deduced the logarithmic series for λ_j. However, he had no data to verify that this series was, in fact, appropriate.

Now suppose we write $\lambda_j = \lambda Q_j$, where $\sum_{j=1}^{\infty} Q_j = 1$; then the general form for $G_L(z)$ becomes

$$G_L(z) = \exp\left\{\lambda[G_Q(z) - 1]\right\} \quad \text{with} \quad G_Q(z) = \sum_{j=1}^{\infty} z^j Q_j,$$

which is (2.12) when $p_n(1)$ is Poisson with mean λ, $z = e^\theta$ and the random variable Y is discrete with probability distribution Q_j. This was mentioned by Ackermann (1939) but has recently been rediscovered (Campagne, 1957; Thyrion, 1960) under such titles as "clustering Poisson" (*par grappes*), "chain reactions," and "cumulative risk." With the foregoing interpretation of Y the right-hand side of (2.10) may be written in the form

$$p_n(1) = \sum_{j=1}^{\infty} e^{-\lambda} \frac{\lambda^j}{j!} Q_n{}^{j*},$$

where $Q_m (m = 1, 2, \ldots)$ is the probability of m deaths occurring in any one accident and

$$Q_n^{j*} = \sum_{m=1}^{n-1} Q_m^{\overline{j-1}*} Q_{n-m}$$

is the probability of n deaths in j accidents. It follows that the logarithmic series of Lüders (1934) is now the logarithmic distribution of the number of deaths per accident, namely,

$$Q_j = \lambda^{-1} \frac{hd^{j-1}}{j(1+d)^j} = \lambda^{-1} \frac{1}{j} \alpha q^j, \qquad j = 1, 2, 3, \ldots,$$

subject to

$$\sum_{j=1}^{\infty} Q_j = -\frac{\alpha}{\lambda} \ln (1-q) = 1.$$

Recently Kupper (1965) reviewed some of the possible forms for Q_j and applied them to the multiple traffic accidents involving bodily injury in the town of Zurich during 1961–1962. Ignoring 4121 accidents which resulted in single injuries, he obtained the results listed in Table 2.4 by using the method of moments to fit G.P. and logarithmic series distributions to the aggregate accidents of the two years.

Table 2.4

Number of Persons Injured	Number of Accidents	Number Expected by Fitted	
		G.P.	Logarithmic Series
2	430	412.6	420.7
3	71	91.9	80.8
4	19	20.4	20.7
5	6	4.6	5.9
6 and more	5[a]	1.3	2.7
	531	530.8	530.8
Goodness of fit		$\chi_2^2 = 10.0$	$\chi_3^2 = 3.5$

[a] Four accidents involving six persons and one involving eleven.

Finally, we mention a generalization of the "proneness" concept introduced by Cresswell & Froggatt (1963). Suppose that every individual passes through a number of spells during which he is capable of having one or more accidents and outside of which no accident can occur. The probability distribution of the number of spells in a given period is assumed to be

Poisson with parameter α and the probability distribution of accidents within a spell is also Poisson but with parameter β.

Now the fmgf of the accidents during any one spell is $e^{\beta s}$ by (2.19) and the fmgf of the number of accidents in k independent spells in $e^{k\beta s}$. Hence the fmgf of Cresswell and Froggatt's so-called "Long" distribution is

$$A_{CF}(s) = \sum_{k=0}^{\infty} e^{-\alpha} \frac{\alpha^k}{k!} e^{k\beta s} = \exp\left[\alpha(e^{\beta s} - 1)\right], \qquad (2.24)$$

from which the individual probabilities can be derived (Irwin, 1964). Now suppose we write an individual's "proneness" to r spells as a discrete distribution with

$$dU(r\beta) = \frac{e^{-\alpha}(\alpha^r/r!)}{1 - e^{-\alpha}}, \qquad r = 1, 2, \ldots. \qquad (2.25)$$

We then note that

$$\int_0^{\infty} dU(r\beta) = (1 - e^{-\alpha})^{-1}\left[e^{-\alpha}\alpha + e^{-\alpha}\frac{\alpha^2}{2!} + e^{-\alpha}\frac{\alpha^3}{3!} + \cdots\right]$$

$$= (1 - e^{-\alpha})^{-1} e^{-\alpha}(e^{\alpha} - 1) = 1,$$

as required, and from (2.20)

$$M_U(s) = \int_0^{\infty} e^{\lambda s}\, dU(\lambda) = (1 - e^{-\alpha})^{-1} \sum_{s=1}^{\infty} e^{r\beta s} e^{-\alpha} \frac{\alpha^r}{r!}$$

$$= (1 - e^{-\alpha})^{-1} e^{-\alpha}[\exp(\alpha e^{\beta s}) - 1] = \exp\left[\alpha(e^{\beta s} - 1)\right]$$

$$= A_{CF}(s).$$

This shows that the Cresswell-Froggatt "Long" distribution is a mixed Poisson with discrete "proneness" distribution (2.25). However, (2.24) is actually the fmgf of the Neyman Type A distribution which he derived (1939) explicitly to cope with "contagion" in populations of insects and bacteria. Here again, then, we find "proneness" and "contagion" inextricably confused.

In general, the applied statistician has preferred to work with the "proneness" or heterogeneity model after ruling out, whenever possible, the existence of "contagion." It is to be noted that if M and N are the random variables that express the number of accidents to an individual in two successive time intervals the proneness model implies that M and N are correlated. The "contagion" model specifies a different correlation structure that is very hard to distinguish from "proneness," even with a two-way frequency table of observations. Let us now review some of the practical applications of these models.

Most of the negative binomial fittings in the actuarial literature are based on the "proneness" or "susceptibility" model. This is true, for example, of

Coppini's (1963) and Chiang's (1965) work on repeated sicknesses and of all five of the articles cited on automobile accidents. With such a model we note that

$$\mathcal{E}(n^{(k)} \mid \lambda) = \sum_{n=0}^{\infty} n^{(k)} e^{-\lambda} \frac{\lambda^n}{n!}, \qquad k = 1, 2, 3, \ldots,$$

$$= \lambda^k \sum_{n=k}^{\infty} e^{-\lambda} \frac{\lambda^{n-k}}{(n-k)!} = \lambda^k$$

and

$$\mathcal{E}\mathcal{E}_{\lambda n}\{n^{(k)} \mid \lambda\} = \mathcal{E}n^{(k)} = \mathcal{E}\Lambda^k. \tag{2.26}$$

In particular,

$$\mathcal{V}(n) = \mathcal{E}n^{(2)} + \mathcal{E}n^{(1)} - \{\mathcal{E}n^{(1)}\}^2 = \mathcal{E}\Lambda^2 + \mathcal{E}(n) - (\mathcal{E}\Lambda)^2$$

$$= \mathcal{E}(n) + \mathcal{V}\Lambda,$$

whereas for Poisson

$$\mathcal{V}(n) = \mathcal{E}(n).$$

Thus, when the variance of a distribution of accidents by number is greater than the mean number of accidents per individual, one possibility is that the individual liabilities, or "proneness" to accident, differ.

If this "proneness" concept—which in automobile insurance may be a proneness of contractholders to make claims rather than of individual drivers to have accidents—is to be useful, the distribution of Λ should remain unchanged in successive periods. If a single observational period extends over t units of time so that an individual's Poisson parameter becomes λt instead of λ, it is easily verified that the foregoing relations become

$$\mathcal{E}(n) = t\mathcal{E}\Lambda \quad \text{and} \quad \mathcal{V}(n) = t\mathcal{E}\Lambda + t^2 \mathcal{V}\Lambda, \tag{2.27}$$

so that the quantity

$$\frac{\mathcal{V}(n) - \mathcal{E}(n)}{\{\mathcal{E}(n)\}^2} = \frac{\mathcal{V}\Lambda}{\{\mathcal{E}\Lambda\}^2}$$

$$= \text{the square of the coefficient of variation of } \Lambda$$

is independent of t. If, therefore, observational values of this ratio are calculated on the same data with different t-values, they should be relatively constant; for example, in a five-year observational period Chambers and Yule (1941) obtained the values of the ratio for accidents to 166 London bus drivers and 101 shipwright apprentices as shown in Table 2.5. The decrease in the cumulative results implies that the coefficient of variation of Λ is diminishing with time. Actually the mean of Λ was also decreasing during the five-year period. Similar methods were utilized by Johnson and

Table 2.5

Years	Bus Drivers		Shipwrights	
	Individual Years	Cumulative from Start	Individual Years	Cumulative from Start
1	0.578	0.578	1.392	1.392
2	0.475	0.525	1.524	1.386
3	0.444	0.500	0.892	1.277
4	0.270	0.452	1.128	1.173
5	0.444	0.456	0.502	1.118

Garwood (1957), and Bailey (1960) had the idea of verifying the supposed homogeneity of a rating class by calculating the (square of the) coefficient of variation of the observed rates within the class and comparing it with a ratio calculated as above.

Note that if we have an observed distribution of claims on N policies during the course of a year we may estimate the first four moments of the distribution of Λ by calculating the factorial moments of n, namely,

$$\frac{1}{N} \sum_{n=1}^{\infty} v_n n^{(k)},$$

where v_n is the number of contractholders with n claims, and by changing them to moments about the mean of Λ (Kendall & Stuart, 1958) we may fit the appropriate Pearson curve (Pearson & Hartley, 1966; Table 43) as an estimate of the distribution of Λ. This method dates back to Newbold (1927), who applied to it 11 sets of industrial accidents. Eight of the resulting Λ distributions were J-shaped (five of them, Type I, and the remainder, Type VI—the form assumed by the F-distribution) but three of them had Λ-moments that were "impossible" for the Pearson family. We note that the (J-shaped) Type III line is the boundary between the Type I and Type VI areas on the (β_1, β_2) chart. Delaporte (1960, 1963), having selected Type III as appropriate (because of the success of the negative binomial), preferred to estimate its parameters by maximum likelihood. The latter article contains eight illustrations of Type III "accident proneness" distributions calculated from French accident records in the years 1958–1959.

Another approach is to try to estimate the d.f. $U(\cdot)$ as a step-function. Huber (1965) has sketched the relevant theory and has mentioned possible approaches without going into detail. Two very different solutions of the problem have been published; Grenander's (1957) utilizes a least-squares technique on a chosen number of discrete probability masses at given intervals δ apart on the λ-scale. On the other hand, Tucker (1963) finds the jump-points of λ as the solution of an algebraic-moment problem.

A highly simplified version of the "proneness" model is to make $U(\cdot)$ a step function with two steps corresponding to good and bad risks, respectively. Such a model was successfully fitted to 23,589 automobile (24–45 hp) accident claims by Tröbliger (1961), who found 94.03 % of the risks to be "normal" with $\lambda = 0.1089$ and 5.97 % to be poor risks with $\lambda = 0.7$. The resulting distribution was as follows:

Number of Claims in Year	Number of Contracts	Expected Number
0	20,592	20,589
1	2,651	2,656
2	297	289
3	41	44
4	7	7
5	—	1
6	1	—
	23,589	23,586
Goodness of fit		$\chi_1^2 = 0.44$

Lundberg (1940) proved that a necessary and sufficient condition for $p_n(t)$ to have the mixed Poisson form

$$p_n(t) = \int_0^\infty e^{-\lambda t} \frac{(\lambda t)^n}{n!} \, dU(\lambda) \tag{2.28}$$

is that the conditional probability of m events during $(0, \tau)$, given that $n \geq m$ events have occurred in the interval $(0, t)$, where $t > \tau$, equals

$$\binom{n}{m} \left(\frac{\tau}{t}\right)^m \left(1 - \frac{\tau}{t}\right)^{n-m}.$$

He applied this result to 1417 sickness indemnity policies issued during 1918–1922 on male lives. The m and n of the foregoing formulas were, respectively, the number of sickness claims occurring in the fourth through the eighth years of an individual policy and those occurring in the fourth through the thirteenth policy years. As tested by chi-squared, eight of nine n-values provided reasonable fits.

When $U(\lambda)$ assumes the particular gamma form specified in the relation following (2.20), and provided that an individual's λ does not change in a period of length s following on the interval $(0, t)$, it is easy to show (Arbous and Kerrich, 1951) that the pgf of the bivariate distribution of N and L, the random variables representing the number of accidents suffered in the two nonoverlapping periods, is

$$G(u, v) \equiv \sum_{n=0}^{\infty} \sum_{l=0}^{\infty} p_{nl}(t, s) u^n v^l$$

$$= (1 + A + B - Au - Bv)^{-m}, \qquad A = as, \quad B = at, \quad (2.29)$$

with the result that

$$p_{nl}(t, s) = \frac{1}{(1 + A + B)^{m+n+l}} \frac{\Gamma(m + n + l)}{\Gamma(m)} \frac{A^n}{n!} \frac{B^l}{l!}. \qquad (2.30)$$

These authors fitted the bivariate distribution (2.30) by the method of moments to the accidents of 122 experienced train shunters in the first six and the following five of 11 observational years (Adelstein, 1952). The fit was judged satisfactory. The same distribution was later applied by Bates and Neyman (1954), who used maximum likelihood estimates of the three parameters, to 10 bivariate accident distributions. Five of the fits were reasonable, but in the remainder the poor fit led to the suspicion that an aberrant subgroup of individuals was present. Recently (2.30) was fitted to four groups of children observed over two consecutive four-year periods, and the numbers of their "medically attended injuries" were in good agreement with the model (Mellinger et al., 1965).

In the meantime Consael (1952) had introduced a more general form for (2.29) in which an additional term in uv appeared. This allows for the possibility that an individual's λ has changed to $\mu \neq \lambda$ in the subsequent period s. Edwards and Gurland (1961) fitted this distribution by moments to two consecutive years' accidents incurred by the 166 London bus drivers mentioned above. In a goodness of fit test of the results $\chi_{31}^2 = 28.63$. Previously Hofmann (1955) had fitted an even more general form of bivariate distribution to work versus nonwork accidents suffered during 1944–1952 by 1196 male transport workers in Zurich. Here $\chi_{32}^2 = 38.5$.

Although these successes with the bivariate versions of the "proneness" model do not "prove" that proneness exists, they indicate that the model is a convenient working hypothesis.

$$* \qquad * \qquad *$$

Commencing with relation (2.18), $p_n(t)$ has referred to n claims made by a single policyholder instead of an aggregate of n claims suffered by a risk business. If we utilize one of the models of the preceding paragraphs to find the d.f. $F(x, t)$ of the aggregate claim made by a single contractholder, the d.f. of the total claims suffered by a risk business composed of N contract-holders with identical distributions $F(\cdot)$ is, say,

$$F_T(x, t) = F^{N*}(x, t). \tag{2.31}$$

If N is large, we could apply the Central Limit theorem and approximate $F_T(\cdot)$ by a Normal d.f., or we could fit one of the Pearson family of curves. There appear to be no numerical examples of this procedure in the literature. From one viewpoint (2.31) suffers from the fact that although t may be made relatively large in determining $F(x, t)$ the "whole portfolio" of size N limits our forecast to a risk business of constant size.

The Distribution $P(y)$

Wholly analogous to the d.f. of an individual claim is the "continuation table," "persistency function," or "reduction factor" of sickness indemnity insurance. Such reduction factors are $1 - P(y)$ in our notation, in which y denotes the length of time since the commencement of an individual's sickness. The earliest attempt to provide a mathematical law for such a distribution was that of Moser (1901) for the sickness fund of the canton of Berne in Switzerland between 1884 and 1893. He graduated the data by the formula

$$1 - P(y) = s^y \exp\{A[(c + y)^{-1} - c^{-1}]\},$$

although, in accordance with actuarial practice, the emphasis in his and the papers mentioned below was on the force of recovery from sickness, namely,

$$\mu(y) = \frac{P'(y)}{1 - P(y)},$$

which is Moser's case assumed the form

$$\mu(y) = a + \frac{b}{(c + y)^2}.$$

Horta & Leão (1948) and Medin (1951) subsequently used the modified formula

$$1 - P(y) = \exp\{A[(c + y)^{-b} - c^{-b}]\},$$

but before this there had been some success in Sweden with the simpler Pareto form

$$1 - P(y) = A(c + y)^{-b}$$

for length of disability (Philipson, 1936) even with $b = 1$ (Askelöf & Cramér, 1932). However, in these last two cases it was not found possible to use the same parameter values for the whole range of y. The well-known Gompertz-Makeham formula

$$\mu(y) = A + Bc^y$$

was used successfully by Gauthier (1958) on the disability durations in excess of a year of French insurance company personnel, and this formula was also used on subclassifications of a more extensive set of data from the same source (Wetzel, 1963). The latter author found it necessary to use the double Gompertz

$$\mu(y) = A + B_1 c_1{}^y + B_2 c_2{}^y$$

for sickness durations of less than a year.

Perhaps the first attempts to fit frequency distributions to sickness costs (as opposed to sickness durations) are those reported by Amoroso (1934). We here have a record of seven different distributions of claims by amount fitted by the lognormal distribution—two of them by the method of moments, the remainder by equating two points on the observed and theoretical d.f.'s. This author also used the lognormal on seven sickness-duration distributions (*loc. cit.*) and eight more may be found in a later article by an author of the same name (1942). Lognormal distributions have also been fitted to distributions of specialty drug costs and of standard drug prescription costs (Cannella, 1963), to workmen's compensation costs (Bühlmann & Hartmann, 1956; Latscha, 1956; Dropkin, 1966), to commercial automobile claims (Henry, 1937; Bailey, 1942–1943); and to fire insurance claim costs (Benckert, 1957, 1963). Not all of the graduations were tested properly for goodness of fit and some of the authors mention the necessity of discarding small claims.

A distribution that has been used successfully on fire insurance (Benckert & Sternberg, 1957) and automobile claims (Benktander, 1963) is the Pareto, namely,

$$1 - P(y) = \left(\frac{y}{y_0}\right)^{-\alpha}, \qquad y_0 < y < \infty, \qquad (2.32)$$

where $\mu(y)$ has the simple form α/y. An awkward feature of this distribution is that the jth moment about the origin

$$\mathcal{E}Y^j \equiv p_j = y_0 \left(\frac{\alpha}{\alpha - j}\right)^{1/j}, \qquad j < \alpha,$$

becomes infinite when $j \geq \alpha$. This means, for example, that the variance of the distribution exists only when $\alpha > 2$. One way of avoiding this difficulty is to set a finite upper limit ω on y as was done in the first of the two articles cited above, in which, in seven successful applications of maximum likelihood estimation, $\hat{\alpha}$ was in the neighborhood of 1.5.

It is interesting to observe that if the parameter α in the d.f. $1 - e^{-\alpha y}$ were a random variable with a gamma distribution having an index m and scale factor c^{-1} [*cp.* the relation following (2.20)] the unconditional d.f. of Y would be given by

$$P(y) = 1 - \left(1 + \frac{y}{c}\right)^{-m}, \qquad 0 < y < \infty$$

(Maguire et al., 1952; Thyrion, 1964), which by linear transformation of y can be put in the form (2.32). Thus, if the individual contracts of a risk business could be regarded as having a negative exponential claim distribution with scales determined by sampling from a gamma distribution, the resulting unconditional distribution of individual claims would be Pareto.

Finally we mention the possibility that $P(y)$ can be graduated by a mixture of negative exponentials (Almer, 1957; Andreasson, 1966; Dropkin, 1966) or, more generally, of gamma distributions (Coppini, 1963). This type of graduation implicitly recognizes that the smaller and larger claims have different *raisons d'être*, whether on economic or moral grounds.

A convenient index to the relative spread of $P(y)$ is provided by

$$(p_2 - p_1^2)^{1/2}/p_1,$$

the coefficient of variation of the distribution. Table 2.6 gives this coefficient for some of the foregoing distributions.

The Convolution-Mixed Distribution $F(x, t)$

An advantage of the model characterized by relation (2.10) is that only the factor $p_n(t)$ depends on t and that, once this probability has been estimated as a function of t, $F(x, t)$ can be calculated for all t-values. On the other hand, a substantial disadvantage is that even for relatively simple mathematical forms of $P(x)$ it is difficult to compute $P^{n*}(x)$ directly and resort must usually be had to the moments of $P(x)$ and its convolutions.

The particular case of (2.10) in which $p_n(t)$ is a Poisson was first established by Lundberg (1919) and became the basis for his "collective theory of risk" which we develop (without using the adjectival prefix) in Chapter 4. We call this distribution the convolution-mixed Poisson (see Haight, 1967, Section 3.6) since Feller's "compound" versus "generalized" terminology (1943) has not been universally accepted, even by its author.

The immediate problem with

$$F(x, t) = \sum_{n=0}^{\infty} e^{-\lambda t} \frac{(\lambda t)^n}{n!} P^{n*}(x) \qquad (2.33)$$

is its numerical calculation for given t and λ. The only explicit results that have been obtained are those for

$$P'(x) = \frac{a^A}{\Gamma(A)} e^{-ax} x^{A-1}, \qquad\qquad a > 0, \quad A > 0,$$

and

$$P'(x) = \frac{B}{\sqrt{\pi}} x^{-3/2} \exp\left(2aB - \frac{B^2}{x} - a^2 x\right), \qquad a > 0, \quad B > 0,$$

and even these are in infinite series form (Hadwiger, 1942).

Table 2.6

Author and Date	Distribution	Coefficient of Variation
Cannella (1963)	Specialty drug costs	0.745
	Standard drug costs	0.720
Latscha (1956)	Workmen's compensation costs, 1949–1951	7.531
Dropkin (1966)	Workmen's permanent disability claim costs, 1963	0.827
Bailey (1942)	Commercial auto property damage, New York state	1.789
Coppini (1963)	Duration of salaried workers' sickness:	
	Men	1.338
	Women	1.312

The particular case $A = 1$, the negative exponential, implies that the d.f. of an individual claim is given by

$$P(y) = 1 - e^{-y}, \qquad 0 \le y < \infty,$$

when $1/a$, the expected value of a claim, is unity. Relation 2.11 then provides

$$P^{2*}(y) = \int_0^y P(y - z)\, dP(z) = \int_0^y \{1 - e^{-(y-z)}\}e^{-z}\, dz$$

$$= 1 - e^{-y}(1 + y)$$

and, by an inductive procedure,

$$P^{n*}(y) = 1 - e^{-y} \sum_{j=0}^{n-1} \frac{y^j}{j!}, \qquad n = 1, 2, 3, \ldots. \tag{2.34}$$

Also, from (2.33) with $\lambda = 1$ without loss of generality

$$F(x, t) = e^{-t} + \sum_{n=1}^{\infty} e^{-t} \frac{t^n}{n!} \left(1 - e^{-x} \sum_{j=0}^{n-1} \frac{x^j}{j!} \right)$$

$$= e^{-t} + (1 - e^{-t}) - e^{-t-x} \sum_{n=1}^{\infty} \frac{t^n}{n!} \sum_{j=1}^{n} \frac{x^{j-1}}{(j-1)!}$$

$$= 1 - e^{-t-x} \sum_{j=1}^{\infty} \frac{x^{j-1}}{(j-1)!} \sum_{n=j}^{\infty} \frac{t^n}{n!}$$

$$= 1 - e^{-x} \sum_{j=1}^{\infty} \frac{x^{j-1}}{(j-1)!} \frac{1}{\Gamma(j)} \int_0^t e^{-s} s^{j-1}\, ds$$

$$= 1 - e^{-x} \int_0^t e^{-s} \sum_{j=0}^{\infty} \frac{(xs)^j}{(j!)^2}\, ds$$

$$= 1 - e^{-x} \int_0^t e^{-s} I_0(2\sqrt{xs})\, ds, \tag{2.35}$$

where $I_0(z) = \sum_{j=0}^{\infty} (z/2)^{2j}/(j!)^2$ is a modified Bessel function which has been tabulated numerically and satisfies the differential equation.

$$zI_0''(z) + I_0'(z) - I_0(z) = 0.$$

This result is due to Ackerman (1939). An application of this exponential assumption to the aggregate claims arising from multiple accidents is to be found in Strickler (1960).

If $p_n(t)$ is of the form (2.18) with

$$U(\lambda) = 1 - e^{-\lambda}, \qquad 0 \le \lambda < \infty,$$

and $P(\cdot)$ is negative exponential with unit mean, it is easy to show that (Lundberg, 1940)

$$F(x, t) = 1 - \frac{t}{1+t} e^{-x/(1+t)}.$$

When $P(x)$ is a mixed exponential (called an "exponential polynomial" by some writers) with, say, only three components, $F(x, t)$ can be calculated directly with good approximation (Almer, 1957; Bohman & Esscher, 1963–1964).

Another possibility is to approximate the observational $P(y)$ by a step function with only a few steps at equidistant intervals of y. This idea was used very successfully by Pentikäinen (1947), who found experimentally that quite large differences in the size and number of steps had a surprisingly small effect on $F(x, t)$. The following lemma of Pesonen (1967) is useful in this connection: if

$$P(y) = \sum_{i=1}^{k} a_i P_i(y), \quad \text{with} \quad \sum_{i=1}^{k} a_i = 1,$$

then

$$F(x, t) = \prod_{i=1}^{k}{}^* F_i(\cdot, a_i t), \qquad (2.36)$$

where the asterisk denotes a convolution product and the suffix i to $F_i(\cdot)$ means that only $P_i(\cdot)$ is utilized; for example, suppose that

$$P_i(y) = P_i^{1*}(y) = \begin{cases} 0, & y < h_i, \\ 1, & y \ge h_i, \end{cases} \qquad (2.37)$$

so that a_i is the probability that $y = h_i$. Then, from (2.11), suppressing i,

$$P^{2*}(y) = \int_0^{\infty} P^{1*}(y - z)\, dP(z) = P^{1*}(y - h) = P(y - h),$$

$$P^{3*}(y) = \int^{\infty} P^{2*}(y - z)\, dP(z) = P^{2*}(y - h) = P(y - 2h),$$

and, inductively,

$$P^{n*}(y) = P(y - \overline{n - 1}h). \qquad (2.38)$$

Hence

$$F_i(y, a_i t) = \sum_{n=0}^{\infty} e^{-a_i t} \frac{(a_i t)^n}{n!} P_i(y - \overline{n-1}h_i)$$

$$= \sum_{n=0}^{[y/h_i]} e^{-a_i t} \frac{(a_i t)^n}{n!}, \qquad (2.39)$$

and, provided k is fairly small, (2.36) can be used to calculate $F(x, t)$ for the limited number of integer x-values possible.

Hovinen (1967) has described a practical Monte Carlo procedure for approximating $F(\cdot)$ by using $k = 3$ in (2.36). The principle consists of drawing a variate from each of the three distributions and adding them together. If this process is repeated many thousands of times, a good approximation to $F(\cdot)$ will be produced. [We note that Marsaglia (1961) has devised and illustrated a method of representing a d.f. as a linear compound of three uniform d.f.'s.]

In Hovinen's method $P_1(\cdot)$ applies to y-values up to an amount ξ such that the resulting $F_1(\cdot)$ is approximately Normal. [Personen (1967) describes two different ways of finding ξ.] A random value of X_1 is then drawn from the Normal distribution with mean $a_1 t p_{11}$ and variance $a_1 t p_{21}$, where

$$p_{i1} = \int_0^\xi y^i \, dP_1(y), \qquad i = 1, 2.$$

The random variable X_2 derives from $P_2(\cdot)$ which applies to y-values between ξ and η, where, for example,

$$1 - P(\eta) = 0.01 [1 - P(\xi)].$$

In order to generate a value of X_2 a number n_2 of claims is first obtained randomly from a Poisson distribution with parameter $a_2 t$. It would be possible but time-consuming to draw n_2 random values of $Y_2(\cdot)$ and obtain x_2 as

$$x_2 = \sum_{j=1}^{n_2} y_2^{(j)}.$$

This direct method, however, is used only to obtain x_3, since the definition of η implies that n_3, a variate from a Poisson with parameter $a_3 t$, will be small. In the case of $F_2(\cdot)$ a set of tables of $P_2^{2^j*}(\cdot)$ $(j = 1, 2, 3, \ldots)$ is prepared by continued use of the relation

$$P_2^{2^{j-1}*}(\cdot) * P_2^{2^{j-1}*}(\cdot) = P_2^{2^j*}(\cdot), \qquad (j = 1, 2, 3, \ldots).$$

Then the randomly drawn number n_2 is written in binary form:

$$n_2 = \sum_{j=0} b_j 2^j,$$

where $b_j = 0$ means that no claim has occurred in the corresponding convolution $P_2^{2^j*}(\cdot)$. A random value of $Y_2^{(2^j)}$ is accordingly obtained from $P_2^{2^j*}(\cdot)$ for every j-value, where $b_j = 1$ and x_2 is given by summing these values, namely,

$$x_2 = \sum_j y_2^{(2^j)}.$$

Finally, the required randomly drawn X-value is obtained as

$$x = x_1 + x_2 + x_3.$$

Because the work of Charlier was well known to Swedish actuaries, it was natural for Lundberg (1919) to seek an approximation to $F(x, t)$ by means of a series development in terms of the Normal d.f. $\Phi(\cdot)$ and its successive derivatives (Cramér, 1946). For this purpose we require the cumulants of $F(\cdot)$, which we may derive from (2.12) as follows:

$$
\begin{aligned}
M_X(\theta) &= \sum_{n=0}^{\infty} p_n(t)[M_Y(\theta)]^n \\
&= \sum_{n=0}^{\infty} e^{-\lambda t} \frac{(\lambda t)^n}{n!} [M_Y(\theta)]^n = \exp\{\lambda t[M_Y(\theta) - 1]\} \\
&= \exp\left(\lambda t \sum_{j=1}^{\infty} \frac{\theta^j}{j!} p_j\right) = \exp\left(\sum_{j=1}^{\infty} \frac{\theta^j}{j!} \kappa_j\right)
\end{aligned}
\tag{2.40}
$$

by definition, where κ_j is the jth cumulant of X (Kendall & Stuart, 1958). Hence

$$\kappa_j = \lambda t p_j, \qquad j = 1, 2, 3, \ldots, \tag{2.41}$$

a relation apparently due to Esscher (1932), although a recursion formula for μ_j', the jth moment of X about zero, had already been given by Lundberg (1919). These moment relations have been rediscovered several times in the last 30 years and ascribed to different authors.

The Edgeworth series to order $(\lambda t)^{-1}$ is then (Cramér, loc. cit.)

$$F(x, \lambda t) \sim \Phi(z) - \frac{1}{3!} \frac{\kappa_3}{\kappa_2^{3/2}} \Phi'''(z) + \frac{1}{4!} \frac{\kappa_4}{\kappa_2^2} \Phi^{(\mathrm{vi})}(z) + \frac{10}{6!} \frac{\kappa_3^2}{\kappa_2^3} \Phi^{(\mathrm{vi})}(z), \tag{2.42}$$

where

$$z = \frac{x - \lambda t p_1}{(\lambda t p_2)^{1/2}}$$

and, in standard error-function notation,

$$\Phi(z) = \frac{1}{\sqrt{2\pi}} \int_{-\infty}^{z} e^{-x^2/2} \, dx.$$

This series approximation, however, has not been applied very frequently in practice because an alternative approach due to Esscher (1932) has been found to be more accurate (Lefèvre, 1952b; Bohman & Esscher, 1963–1964).

Actually Esscher's (1932, 1963) method is a modification of the so-called saddlepoint approximation of an integral, which in this case is the moment-generating function (i.e., the two-sided Laplace transform with negative argument) of the density function of X. In our development of this approximation we follow Daniels (1954).

Let us write

$$\frac{\partial F(x, t)}{\partial x} = f(x),$$

which implies that $P(\cdot)$ is continuous at the point x, and put

$$\phi(s) = \int_{-\infty}^{\infty} e^{sz} f(z) \, dz \equiv e^{K(s)}, \tag{2.43}$$

where it is convenient to extend the range of integration to the negative axis even though $f(\cdot)$ may be zero for negative arguments. The integral in (2.43) is supposed to exist for a certain range of $\Re s$ and if u assumes a value within this range the inversion of (2.43) may be written

$$f(x) = \frac{1}{2\pi i} \int_{u-i\infty}^{u+i\infty} e^{-sx} \phi(x) \, ds, \tag{2.44}$$

the integration being effected along a line parallel to the imaginary axis.

Now writing $s = u + iv$

$$e^{-sx} \phi(s) = e^{-(u+iv)x} \phi(u + iv) = e^{K(u)-ux} \exp\left[-ivx + K(u + iv) - K(u)\right]$$

$$= e^{K(u)-ux} \exp\left[-ivx + \sum_{j=1}^{\infty} \frac{(iv)^j}{j!} K^{(j)}(u)\right].$$

Keeping u fixed and applying the inversion formula (2.44), we obtain

$$f(x) = e^{K(u)-ux} \frac{1}{2\pi} \int_{-\infty}^{\infty} \exp\left\{[K'(u) - x]iv + \sum_{j=2}^{\infty} \frac{(iv)^j}{j!} K^{(j)}(u)\right\} dv$$

$$= \frac{e^{K(u)-ux}}{2\pi K''^{1/2}(u)} \int_{-\infty}^{\infty} \exp\left[\frac{K'(u) - x}{K''^{1/2}(u)} iw + \sum_{j=2}^{\infty} \frac{(iw)^j}{j!} \frac{K^{(j)}(u)}{K''^{j/2}(u)}\right] dw$$

$$= \frac{e^{K(u)-ux-\xi^2/2}}{[2\pi K''(u)]^{1/2}} \frac{1}{\sqrt{2\pi}} \int_{-\infty}^{\infty} e^{-1/2(w-i\xi)^2} \exp\left[\sum_{j=3}^{\infty} \frac{(iw)^j}{j!} \lambda_j(u)\right] dw,$$

where

$$\xi = \frac{K'(u) - x}{K''^{\frac{1}{2}}(u)} \quad \text{and} \quad \lambda_j(u) = \frac{K^{(j)}(u)}{K''^{j/2}(u)}.$$

On expanding the second exponential in the integrand, changing the variable to $y = w - i\xi$, and integrating term by term we get

$$f(x) = \frac{e^{K(u) - ux - \xi^2/2}}{[2\pi K''(u)]^{\frac{1}{2}}} \left\{ 1 - \frac{1}{3!} \lambda_3(u) H_3(\xi) \right.$$
$$\left. + \left[\frac{1}{4!} \lambda_4(u) H_4(\xi) + \frac{10}{6!} \lambda_3^2(u) H_6(\xi) - \cdots \right] \right\}, \quad (2.45)$$

where $H_n(\xi)$ is the nth degree Hermite polynomial in ξ (Cramér, 1946).

Let us first take $u = 0$ in (2.45). We have

$$K'(u) = \frac{d}{du} \ln \phi(u) = \frac{\phi'(u)}{\phi(u)}$$

and

$$K''(u) = \frac{\phi(u) \phi''(u) - \phi'^2(u)}{\phi^2(u)}.$$

Since

$$\phi^{(k)}(u) = \int_{-\infty}^{\infty} e^{uz} z^k f(z) \, dz, \qquad k = 1, 2, 3, \dots,$$

$K'(0)$ and $K''(0)$ are, respectively, the mean and variance of the random variable whose density function is $f(\cdot)$; ξ is thus minus the standardized variate corresponding to $X = x$, and, making use of the relation (Cramér, 1946),

$$\Phi^{(n+1)}(\xi) = (-1)^n H_n(\xi) \Phi'(\xi),$$

we see that (2.45) reduces to (2.42) when $u = 0$.

On the other hand, when $\xi = 0$, the parameter u has to be determined from

$$K'(u) = \frac{\phi'(u)}{\phi(u)} = x. \quad (2.46)$$

Noting that

$$K''(u) = \int_{-\infty}^{\infty} e^{ux} f(x) \, dx \int_{-\infty}^{\infty} (x - y)^2 e^{uy} f(y) \frac{dy}{2\phi''(u)},$$

as may be verified by expanding the squared term in the integrand and identifying $\phi(\cdot)$ and its derivatives, we see that $K'(u)$ is always increasing so that (2.46) has at most one real root. By rewriting (2.46) as a single integral equated to zero we find that this root always exists (Jánossy, 1965).

Further (Cramér, 1946)

$$H_{2j}(0) = \frac{(-1)^j(2j)!}{j!2^j}$$

$$H_{2j+1}(0) = 0,$$

and we obtain from (2.45) for the case in which $\xi = 0$ and u is given by (2.46)

$$f(x) = \frac{e^{K(u)-uK'(u)}}{[2\pi K''(u)]^{1/2}} \{1 + [\tfrac{1}{8}\lambda_4(u) - \tfrac{5}{24}\lambda_3^2(u)] - \cdots\}. \tag{2.47}$$

The factor preceding the brace is called the first saddlepoint approximation, and the whole expression on the right of (2.47), without inclusion of further terms in the braces,* is known as the second saddlepoint approximation (Jánossy, 1965). It it is assumed that

$$\frac{K^{(j)}(u)}{K''(u)} = O(1), \qquad j = 2, 3, \ldots,$$

the successive terms of the series in braces in (2.45) are of orders $(K''(u))^{-l}$ $\times (l = 0, 1, 2, \ldots)$. The two saddlepoint approximations of (2.47) are likely to be more successful the larger the $K''(u)$.

The foregoing approximate method applies quite generally. In the particular case in which $F(\cdot)$ is given by (2.33) and its mgf and cumulants by (2.40) and (2.41), respectively, (2.46) becomes

$$xe^{\lambda t[M_Y(u)-1]} = e^{\lambda t[M_Y(u)-1]}\lambda t \int_0^\infty e^{uy}y\, dP(y);$$

that is,

$$x = \lambda t \int_0^\infty e^{uy}y\, dP(y). \tag{2.48}$$

The function on the right of this equation assumes the value $\lambda t p_1$ when $u = 0$, and this implies that for the equation to be satisfied for $x > 0$

$$u \lessgtr 0 \quad \text{according} \quad \text{as } x \gtrless \lambda t p_1.$$

Confining ourselves to the first saddlepoint approximation, we note that

$$K(u) = \lambda t[M_Y(u) - 1],$$

$$K'(u) = \lambda t\, M_Y'(u) = \lambda t \int_0^\infty e^{uy}y\, dP(y) = x,$$

---

* The first of these is $(1/2^3 6!)(-120\lambda_6 + 525\lambda_4^2 + 840\lambda_3\lambda_5 - 3150\lambda_3^2\lambda_4 + 1925\lambda_3^4)$.

and

$$K''(u) = \lambda t\, M''_Y(u) = \lambda t \int_0^\infty e^{uy} y^2\, dP(y).$$

These values may be calculated (exactly or approximately) once u has been determined from (2.48) and the result substituted into the leading factor on the right of (2.47). In fact, this procedure was suggested by Pesonen (1964) as a more accurate version of the Esscher formula.

Esscher's approximations result from integrating (2.45), with u held constant to obtain $F(x)$, and from applying relation (2.48) to determine u only at the extremity of the range of integration. This means that ξ is a variable that only becomes zero at that extremity and that the first approximation is

$$F(x) \sim \frac{e^{K(u)}}{[2\pi\, K''(u)]^{1/2}} \int_{-\infty}^x \exp\left[\frac{-uz - (K'(u) - z)^2}{2K''(u)}\right] dz$$

$$= \frac{e^{K(u) - u\, K'(u)}}{\sqrt{2\pi}} \int_{-\infty}^0 \exp\left[\frac{-u\sqrt{K''(u)}\, y - y^2}{2}\right] dy, \qquad y = \frac{z - K'(u)}{\sqrt{K''(u)}},$$

where u is the solution of (2.46) and is assumed to be negative, since y is then negative as required. When a second approximation is needed for $F(x)$, the term in $\lambda_3(u)$ in (2.45) does not vanish; even if the loss of accuracy in utilizing a single u-value over the whole range of integration in $F(x)$ is small, we still need to use the second Esscher approximation to obtain results as good as those of the first saddlepoint approximation.

The two-term Esscher formulas which emerge from the foregoing development may be written

$$F(x, t) \sim e^{K(u) - ux}\left[A_0(-\zeta) + \frac{\bar\mu_3}{3!\bar\mu_2^{3/2}} A_3(-\zeta)\right], \qquad x < \lambda t p_1,$$

$$1 - F(x, t) \sim e^{K(u) - ux}\left[A_0(\zeta) - \frac{\bar\mu_3}{3!\bar\mu_2^{3/2}} \bar A_3(\zeta)\right], \qquad x > \lambda t p_1, \quad (2.49)$$

where u is the solution of (2.48) and

$$\zeta = u\sqrt{\bar\mu_2}, \; \phi(u)\,\bar\mu_j = \int_0^\infty (z - x)^j e^{uz}\, dF(z, t),$$

$$A_0(\zeta) = \int_0^\infty e^{-\zeta z}\, d\Phi(z) \quad \text{and} \quad A_3(\zeta) = \int_0^\infty e^{-\zeta z}\, d\Phi'''(z).$$

Even when the functions $A_0(\cdot)$ and $A_3(\cdot)$ are available in tabular form (Lefèvre, 1952a), these expressions are still not easy to evaluate. Before providing numerical examples, we shall summarize some of the generalizations of (2.33) that have been proposed.

An important theoretical result (Thyrion, 1959) is that if $F(x, t)$ is to be a convolution-mixed Poisson then $p_n(t)$ of (2.10) must be of the form

$$p_n(t) = \sum_{m=0}^{\infty} e^{-\lambda t} \frac{(\lambda t)^m}{m!} Q_n^{m*} \qquad (2.50)$$

already encountered in our discussion of multiple deaths as a result of an accident. We saw there that for $p_n(t)$ to be the negative binomial (2.17) we had to have

$$Q_j = \frac{\alpha}{\lambda t} \frac{1}{j} q^j \quad \text{with} \quad \sum_{j=1}^{\infty} Q_j = 1.$$

Hence

$$1 - q = e^{-\lambda t/\alpha} \qquad (2.51)$$

and

$$Q_j = \frac{\alpha}{\lambda t} \frac{1}{j} (1 - e^{-\lambda t/\alpha}) \qquad j = 1, 2, 3, \ldots.$$

Now on substituting (2.50) into (2.10)

$$F(x, t) = \sum_{n=0}^{\infty} P^{n*}(x) \sum_{m=0}^{\infty} e^{-\lambda t} \frac{(\lambda t)^m}{m!} Q_n^{m*}$$

$$= \sum_{m=0}^{\infty} e^{-\lambda t} \frac{(\lambda t)^m}{m!} \left[\sum_{n=0}^{\infty} P^{n*}(x) Q_n^{m*} \right].$$

This shows that the form (2.33) still applies, provided we replace $P(x)$ with

$$S(x) = \sum_{n=0}^{\infty} P^{n*}(x) Q_n. \qquad (2.52)$$

The mth convolution of $S(x)$ is then the quantity in brackets above. In the particular case in which $p_n(t)$ is negative binomial Q_n in (2.52) is given by (2.51) (Ammeter, 1948).

From a numerical viewpoint (2.52) is not easily evaluated, still less its convolutions, and it is usual to work from the moments of $F(x, t)$ calculated directly from (2.12). General formulas for the first four cumulants (seminvariants) are found in Lundberg (1940), who derived the following asymptotic result for large t:

$$F(tz, t) \sim U\left(\frac{z}{p_1}\right) \quad \text{with} \quad \mathcal{E}\Lambda = 1$$

In the particular case in which $p_n(t)$ is negative binomial

$$M_X(\theta) = \sum_{n=0}^{\infty} \binom{\alpha + n - 1}{n} (1 - q)^\alpha q^n [M_Y(\theta)]^n$$

$$= (1 - q)^\alpha [1 - q M_Y(\theta)]^\alpha = (1 - q)^\alpha \left(1 - q \sum_{j=0}^{\infty} \frac{\theta^j}{j!} p_j\right)^\alpha$$

and the successive moments about zero of the random variable X appear as the coefficients of $\theta^k (k = 1, 2, 3, \ldots)$ in the expansion of the factor last written. In particular, when we convert the origin to the mean $\alpha q/(1 - q)$ (which is the λt of Poisson) and take $p_1 = 1$, the second and third cumulants of X are

$$\kappa_2 = \alpha \frac{q}{1 - q} p_2 + \alpha \left(\frac{q}{1 - q} \right)^2,$$

$$\kappa_3 = \alpha \frac{q}{1 - q} p_3 + 3\alpha \left(\frac{q}{1 - q} \right)^2 p_2 + 2\alpha \left(\frac{q}{1 - q} \right)^3.$$

These cumulants can be used in the series approximations (2.42) and (2.49), which were derived without utilizing the Poisson assumption explicitly.

We mention in conclusion that the case in which $p_n(t)$ is approximated by a Normal and $P(y)$ by a gamma distribution has been elegantly treated by Bühlmann (1958).

The numerical comparisons of $1 - F(x, t)$ in Tables 2.7 and 2.8 are taken from the work by Bohman & Esscher (1963–1964) already referred to. We restrict ourselves to two assumptions for $P(y)$, namely, the negative exponential with $p_2 = 2!$ and $p_3 = 3!$ and a distribution representative of those occurring in nonindustrial fire insurance with $p_2 = 47.5854$ and $p_3 = 12,600.1$. In both cases $p_1 = 1$ and the "exact" values were computed by a method of inversion of characteristic functions developed by Bohman (1963).* The negative binomial utilized had $\alpha = 20$ ($\alpha \to \infty$ corresponds to Poisson) and the mean of X is indicated in the tables. The column heading "Lundberg" stands for (2.49) with the omission of the term in $A_3(u)$. For the well-behaved negative exponential (2.49) is an excellent approximation even when the mean is as low as 100, but for the long-tailed fire distribution the mean has to be at least 1000 before (2.49) can be regarded as adequate. The Lundberg approximation is never very good for the fire insurance case, but it is reasonable for a mean of 10,000 in the negative exponential. In general, the negative binomial does not seem to make the fit much worse than it is in the case of Poisson.

Bohman & Esscher (*loc. cit.*) found Pearson's Type III curve with three parameters (i.e., with the start of the curve determined from the third moment) to be rather poor for values of the mean below 1000 in the fire insurance case; but they did not attempt a four-parameter Pearson curve (Elderton, 1938) nor did they check what would happen if higher moments (cumulants) than the third were used in (2.42). Perhaps with more or less "exact" methods available on a large computer these approximate desk-machine methods are becoming a thing of the past.

---

* Details are provided in Appendix B.

Table 2.7   Negative Exponential

$p_n(t)$	$\dfrac{x - \text{mean}}{\kappa_2^{1/2}}$	"Exact"	Lundberg	(2.49)
		Mean = 100		
Negative binomial	0	0.4691	0.5000	0.4691
	1	0.1569	0.1695	0.1565
	2	0.0337	0.0361	0.0336
	3	0.0051	0.0054	0.0051
	4	0.0006	0.0006	0.0006
Poisson	0	0.4859	0.5000	0.4859
	1	0.1584	0.1639	0.1583
	2	0.0282	0.0290	0.0281
	3	0.0028	0.0029	0.0028
	4	0.0002	0.0002	0.0002
		Mean = 1000		
Negative binomial	0	0.4702	0.5000	0.4702
	1	0.1569	0.1691	0.1565
	2	0.0334	0.0356	0.0333
	3	0.0049	0.0052	0.0049
	4	0.0005	0.0006	0.0005
Poisson	0	0.4955	0.5000	0.4955
	1	0.1586	0.1604	0.1586
	2	0.0245	0.0248	0.0245
	3	0.0018	0.0018	0.0018
	4	0.0001	0.0001	0.0001
		Mean = 10,000		
Negative binomial	0	0.4703	0.5000	0.4703
	1	0.1569	0.1691	0.1565
	2	0.0334	0.0356	0.0333
	3	0.0049	0.0052	0.0049
	4	0.0005	0.0006	0.0005
Poisson	0	0.4986	0.5000	0.4986
	1	0.1587	0.1592	0.1587
	2	0.0233	0.0234	0.0233
	3	0.0015	0.0015	0.0015
	4	0.0000	0.0000	0.0000

Table 2.8 Nonindustrial Fire Insurance

$p_n(t)$	$\dfrac{x - \text{mean}}{\kappa_2^{1/2}}$	"Exact"	Lundberg	(2.49)
		Mean = 100		
Negative binomial	0	0.3801	0.5000	0.2670
	1	0.1006	0.2118	0.1148
	2	0.0352	0.0973	0.0665
	3	0.0168	0.0469	0.0358
	4	0.0086	0.0230	0.0187
Poisson	0	0.3743	0.5000	0.2448
	1	0.0947	0.2150	0.1147
	2	0.0345	0.1012	0.0688
	3	0.0171	0.0500	0.0381
	4	0.0089	0.0251	0.0205
		Mean = 1000		
Negative binomial	0	0.4476	0.5000	0.4460
	1	0.1502	0.1761	0.1502
	2	0.0397	0.0457	0.0400
	3	0.0089	0.0100	0.0090
	4	0.0018	0.0020	0.0018
Poisson	0	0.4265	0.5000	0.4193
	1	0.1364	0.1832	0.1463
	2	0.0452	0.0557	0.0472
	3	0.0140	0.0155	0.0138
	4	0.0035	0.0040	0.0037
		Mean = 10,000		
Negative binomial	0	0.4697	0.5000	0.4697
	1	0.1569	0.1693	0.1565
	2	0.0335	0.0358	0.0335
	3	0.0050	0.0053	0.0050
	4	0.0006	0.0006	0.0006
Poisson	0	0.4742	0.5000	0.4745
	1	0.1567	0.1678	0.1565
	2	0.0321	0.0342	0.0322
	3	0.0045	0.0047	0.0045
	4	0.0004	0.0005	0.0004

REFERENCES

Ackermann, W. -G. (1939). "Eine Erweiterung des Poissonschen Grenzwertsatzes und ihre Anwendung auf die Risikoprobleme in der Sachversicherung." *Schr. Math. Inst. Inst. Angew. Math. Univ. Berlin*, **4**, 211–255.

Adelstein, A. M. (1952). "Accident proneness: A criticism of the concept based upon an analysis of shunters' accidents." *J. Roy. Statist. Soc. A.*, **115**, 354–410.

Aitken, A. C. (1957). *Statistical Mathematics*. Oliver and Boyd, Edinburgh.

Almer, B. (1957). "Risk analysis in theory and practical statistics." *Trans. XV Intern. Cong. Actu.*, New York, **2**, 314–349.

Ammeter, H. (1948). "A generalization of the collective theory of risk in regard to fluctuating basic-probabilities." *Skand. Aktu. Tidskr.* **31**, 171–198.

Amoroso, E. (1942). "Nuove ricerche intorno alla distribuzione delle malattie per durata." *Atti. Ist. Naz. Assic.*, **14**, 185–202

Amoroso, L. (1934). "La rappresentazione analitica delle curve di frequenza nei sinistri di infortuni e di responsabilità civile." *Atti X Cong. Intern. Attu.*, **3**, 458–472.

Andreasson, G. (1966). "Distribution free approximations in applied risk theory." *Astin Bull.*, **4**, 11–18.

Anscombe, F. J. (1950). "Sampling theory of the negative binomial and logarithmic series distributions." *Biometrika*, **37**, 358–382.

Arbous, A. G., and J. E. Kerrich (1951). "Accident statistics and the concept of accident proneness." *Biometrics*, **7**, 340–432.

Askelöf, T., and H. Cramér (Eds.) (1932). *Statistiska Undersökningar samt Kostnadsberäkningar* M.M.II. 1928 Ars Pensions-försäkringskommitteé och Organizationssakkunniga, Stockholm.

Bailey, A. L. (1942–1943) "Sampling theory in casualty insurance." *Proc. Causalty Actu. Soc.*, **29**, 50–95; **30**, 31–65.

Bailey, R. A. (1960). "Any room left for skimming the cream?" *Proc. Casualty Actu. Soc.*, **47**, 30–36

Bates, G. E., and J. Neyman (1954). "Contributions to the theory of accident proneness." *U. Calif. Publ. Statist.*, **1**, 215–275.

Benckert, L.-G. (1957). "The premium in insurance against loss of profit due to fire as a function of the period of indemnity." *Trans. XV Intern. Cong. Actu.*, New York, **2**, 297–305.

———— (1963). "The lognormal model for the distribution of one claim." *Astin Bull.*, **2**, 9–23.

Benckert, L.-G., and I. Sternberg (1957). "An attempt to find an expression for the distribution of fire damage amount." *Trans. XV Intern. Cong. Actu.*, New York, **2**, 288–294.

Benktander, G. (1963). "Notes sur la distribution conditionée du montant d'un sinistre par rapport à l'hypothèse qu'il y a eu un sinistre dans l'assurance automobile." *Astin Bull.*, **2**, 24–29.

Berger, A. (1939). *Mathematik der Lebensversicherung*. Springer, Vienna.

Bohlmann, G. (translated and revised by H. P. du Motel), (1909). "Technique de l'assurance sur la vie." *Encycl. Sci. Math.* I 4 (3), 491–590.

Bohman, H. (1963). "To compute the distribution function when the characteristic function is known." *Skand. Actu. Tidskr.*, **46**, 41–46.

Bohman, H., and F. Esscher (1963–1964). "Studies in risk theory with numerical illustrations concerning distribution functions and stop loss premiums." *Skand. Aktu. Tidskr.* **46**, 173–225; **47**, 1–40.

Bühlmann, H. (1958). "Ein theoretischen Beitrag zur statistischen Erfassung der Gesamt-betriebsunfallkosten." *Mitt. Verein. Schweiz. Versich. Mathr.*, **58**, 53–65.

Bühlmann, H., and W. Hartmann (1956) "Änderungen in der Grundgesamtheit der Betriebsunfallkosten." *Mitt. Verein. Schweiz. Versich. Mathr.*, **56**, 303–320.

Camp, K. (1950). *The Whittaker-Henderson Graduation Process.* The Author, New York.

Campagne, C. (1957). "The influence of chain reactions on the loss distribution function." *Trans. XV Intern. Cong. Actu.*, New York, **2**, 248–261.

Campagne, C., B. H. De Jongh, & J. N. Smit (1947). *Bijdrage tot de Wiskundige Theorie van de Bedrijfsreserve en het Eigenbehoud in de Brandverzekering.* Rijksuitgeverij, The Hague.

Cannella, S. (1963). "Variation de la prime d'assurance de l'assistance pharmaceutique en fonction de la participation de l'assuré au coût de l'assistance." *Astin Bull.*, **2**, 30–44.

Chambers, E. G., and G. U. Yule (1941). "Theory and observation in the investigation of accident causation." *J. Roy. Statist. Soc. Suppl.*, **7**, 89–109.

Chiang, C. L. (1965). "An index of health: mathematical models." *Nat. Center Health Statist.*, Ser. 2, No. 5. Washington, D.C.

Consael, R. (1952). "Sur les processus composés de Poisson à deux variables aléatoires." *Acad. Roy. Belg. Cl. Sci. Mém. Coll. in 8°*, **27**, fasc. 6.

Coppini, M. A. (1963). "À propos de la distribution des cas de maladie entre les assurés et par rapport à la durée." *Astin Bull.*, **2**, 45–61.

Cramér, H. (1946). *Mathematical Methods of Statistics.* Princeton University Press, Princeton, N.J.

Cresswell, W. L., and P. Froggatt (1963). *The Causation of Bus Driver Accidents: An Epidemiological Study.* Oxford University Press, London.

Daniels, H. E. (1954). "Saddlepoint approximations in statistics." *Ann. Math. Statist.*, **25**, 631–650.

Davis, H. T. (1963). *Tables of the Mathematical Functions*, Vol. I. Principia Press, San Antonio, Texas.

Delaporte, P. (1960). "Un problème de tarification de l'assurance accidents d'automobiles examiné par la statistique mathématique." *C. R. XVI Cong. Intern. Actu.*, Brussels, **2**, 121–130.

——— (1963). "L'estimation statistique progressive de risque individuel d'accident et la tarification de l'assurance automobile." *Bull. Inst. Intern. Statist.*, Ottawa, **40**, 1, 275–284.

De Morgan, A. (1837). "On a question in the theory of probabilities." *Trans. Camb. Phil. Soc.* **6**, 423–430.

Derron, M. (1962). "Mathematische Probleme der Automobilversicherung." *Mitt. Verein. Schweiz. Versich. Mathr.*, **62**, 103–124.

Dropkin, L. B. (1959). "Some considerations on automobile rating systems utilizing individual driving records." *Proc. Casualty Actu. Soc.*, **46**, 165–176.

——— (1966). "Loss distributions of a single claim." *The Mathematical Theory of Risk.* Casualty Actu. Soc., New York.

Edwards, C. B., and J. Gurland (1961). "A class of distributions applicable to accidents." *J. Amer. Statist. Assoc.*, **56**, 503–517.

Eggenberger, F. (1924). "Die Warhscheinlichkeitsansteckung." *Mitt. Verein. Schweiz. Versich. Mathr.*, **19**, 31–143.

Elderton, W. P. (1938). *Frequency Curves and Correlation.* Cambridge University Press, Cambridge.

Esscher, F. (1932), "On the probability function in the collective theory of risk." *Skand. Aktu. Tidskr.*, **15**, 175–195.

——— (1963). "On approximate computation of distribution functions when the corresponding characteristic functions are known." *Skand. Aktu. Tidskr.*, **46**, 78–86.

Feller, W. (1943). "On a general class of 'contagious' distributions." *Ann. Math. Statist.*, **14**, 389–400.

Gauthier, M. (1958). "De l'ajustement analytique d'une loi de survie en état d'invalidité." *Bull. Trim. Inst. Actu. Franç.*, **57**, 107–120.

Greenwood, M., and G. U. Yule (1920). "An inquiry into the nature of frequency distributions representative of multiple happenings with particular reference to the occurrence of multiple attacks of disease or of repeated accidents." *J. Roy. Statist. Soc.* **83**, 255–279.

Grenander, U. (1957). "On heterogeneity in non-life insurance." *Skand. Aktu. Tidskr.*, **40**, 71–84, 153–179.

Hadwiger, H. (1942). "Wahl einer Näherungsfunktion für Verteilungen auf Grund einer Funktionalgleichung." *Bl. Versich. Math.*, **5**, 345–352.

Haight, F. A. (1967). *Handbook of the Poisson Distribution*. Wiley, New York.

Henry, M. (1937). "Étude sur le coût moyen des sinistres en responsabilité civile automobile." *Bull. Trim. Inst. Actu. Franç.*, **43**, 113–178.

Hesselberg, I. (1937). "La réassurance des excédents de sinistres en assurances sur la vie." *C.R. XI Cong. Intern. Actu.*, Paris, **1**, 417–432.

Hewitt, C. C., Jr. (1967). "Loss ratio distributions: A Model." *Proc. Casualty Actu. Soc.*, **54**, 70–93.

Hickman, J. (1964). "A statistical approach to premiums and reserves in multiple decrement theory." *Trans. Soc. Actu.*, **16**, 1–16.

Hofmann, M. (1955). "Über zusammengesetzte Poisson-Prozesse und ihre Anwendungen in der Unfallversicherung." *Mitt. Verein Schweiz. Versich. Mathr.*, **55**, 499–575.

Horta, E., and A. Leão (1948). "Ensaio de uma tábua de morbidez portuguese." *Bol. Inst. Actu. Portug.* **3**, 55–69.

Hovinen, E. (1967). "A procedure to compute values of the generalized Poisson function." *Astin Bull.*, **4**, 129–135.

Huber, J. (1965). "Neuere Fortschritte in der Theorie der Bayesschen Statistik." *Mitt. Verein. Schweiz. Versich. Mathr.*, **65**, 163–171.

Irwin, J. O. (1964). "The personal factor in accidents." *J. Roy. Statist. Soc. A*, **127**, 438–451.

Jánossy, L. (1965). *Theory and Practice of the Evaluation of Measurements*. Oxford University Press, London.

Johnson, N. L., and F. Garwood (1957). "An analysis of the claim records of a motor insurance company." *J. Inst. Actu.*, **83**, 277–294.

Kendall, M. G., and A. Stuart (1958). *The Advanced Theory of Statistics: Vol. 1, Distribution Theory*. Griffin, London.

Kupper, J. (1960, 1962). "Wahrscheinlichkeitstheoretische Modelle in der Schadenversicherung." *Bl. Deuts. Gesell. Versich. Math.*, **5**, 451–503; **6**, 95–130.

———— (1965). "Some aspects of cumulative risk." *Astin Bull.*, **3**, 85–103.

Latscha, R. (1956). "Zur Anwendung der kollektiven Risikotheorie in der schweizerischen obligatorischen Unfallversicherung." *Mitt. Verein. Schweiz. Versich. Mathr.*, **56**, 275–302.

Lefèvre, J. (1952a). "Application de la théorie collective du risque à la réassurance 'Excess-loss'." *Skand. Actu. Tidskr.*, **35**, 160–187.

———— (1952b). "Formules de calcul des fonctions d'Esscher an théorie collective du risque." *Bull. Ass. Roy. Actu. Belges*, **56**, 27–33.

Lüders, R. (1934). "Die Statistik der seltenen Ereignisse. "*Biometrika*, **26**, 108–128.

Lundberg, F. (1903). I. *Approximerad Framställning af Sannolikhetsfunktionen*. II. *Återförsäkring af Kollektivrisker*. Almqvist and Wiksells, Uppsala.

———— (1919) "Teori för riskmassor." *Försäkringsbolags Fondbildning och Riskutjämning*, II. Beckman, Stockholm.

Lundberg, O. (1940). *On Random Processes and their Application to Sickness and Accident Statistics*. Almqvist and Wiksells, Uppsala.

McKendrick, A. G. (1926). "Application of mathematics to medical problems." *Proc. Edinb. Math. Soc.*, **44**, 98–130.

Maguire, R. A., E. S. Pearson, and A. H. A. Wynn (1952). "The time intervals between industrial accidents." *Biometrika*, **39**, 168–180.

Marsaglia, G. (1961). "Expressing a random variable in terms of uniform random variables." *Ann. Math. Statist.*, **32**, 894–898.

Medin, K. (1951). "A function for smoothing tables of the duration of sickness." *Skand. Aktu. Tidskr.*, **34**, 45–52.

Mellinger, G. D., D. L. Sylwester, W. R. Gaffey, and D. I. Manheimer (1965). "A mathematical model with applications to a study of accident repeatedness among children." *J. Amer. Statist. Assoc.*, **60**, 1046–1059.

Moser, C. (1901). "Communication touchant une table de morbidité." *C.R. III Cong. Intern. Actu.*, Paris, 662–664.

Newbold, E. M. (1927). "Practical applications of the statistics of repeated events particularly to industrial accidents." *J. Roy. Statist. Soc.* **90**, 487–547.

Neyman, J. (1939). "On a new class of 'contagious' distributions, applicable in entomology and bacteriology." *Ann. Math. Statist.* **10**, 35–57.

Pearson, E. S., and H. O. Hartley (Eds.) (1966). *Biometrika Tables for Statisticians*, Vol. I. Cambridge University Press, Cambridge.

Pentikäinen, T. (1947). "Einige numerische Untersuchungen über das risikotheoretische Verhalten von Sterbekassen." *Skand. Aktu. Tidskr.*, **30**, 75–87.

Pesonen, E. (1964). "A modification of the Esscher method." *Skand Aktu. Tidskr.* **47**, 160–163.

——— (1967). "On the calculation of the generalized Poisson function." *Astin Bull.*, **4**, 120–128.

Philipson, C. (1936). "Statistical study of waivers of premiums on disablement." *Skand. Aktu. Tidskr.*, **19**, 71–107, 161–193.

Pizzetti, E. (1964). "La riassicurazione non proporzionale applicata al ramo vita." *Gior. Ist. Ital. Attu.*, **27**, 122–139.

Pollaczek-Geiringer, H. (1928). "Über die Poissonsche Verteilung und die Entwicklung willkürlichen Verteilungen." *Z. Angew. Math. Mech.*, **8**, 292–308.

Ramel, M. (1960). "Tarification d'un traité d'excédent de pourcentage de sinistres." *C.R. XVI Cong. Intern. Actu.*, Brussels, **1**, 549–561.

Riebesell, P. (1936). "Einführung in die Sachversicherungsmathematik." *Veröff. Deuts. Vereins Versich. Wissens.*, **56**, 1–90.

Roberts, G. E., and H. Kaufman (1966). *Table of Laplace Transforms*. Saunders, Philadelphia, Pa.

Rossman, G. (1938) "Ajustement des écarts en assurance-grêle. *Bull. Trim. Inst. Actu. Franç.*, **44**, 75–78.

Seal, H. L. (1947). "A probability distribution of deaths at age x when policies are counted instead of lives." *Skand. Aktu. Tidskr.*, **30**, 18–43.

——— (1949). "Discrete random processes." *J. Inst. Actu. Stud. Soc.* **8**, 203–209.

——— (1951). Correspondence. *J. Inst. Actu.* **77**, 490–492.

——— (1953). "The mathematical risk of lump-sum death benefits in a trusteed pension plan." *Trans. Soc. Actu.*, **5**, 135–142.

Sergowskij, N. (1937) "Schwankungsintensität der Schadensquote und Schwankungsreserve in der Feuerversicherung." *C.R. XI Cong. Intern. Actu.*, Paris, **2**, 523–559.

Simon, L. J. (1961). "Fitting negative binomial distributions by the method of maximum likelihood." *Proc. Casualty Actu. Soc.*, **48**, 45–53.

——— (1965). "The 1965 Table *M*." *Proc. Casualty Actu. Soc.*, **52**, 1–51.

Stone, D. G. (1948). "Mortality fluctuation in a small self-insured pension plan." *Trans. Actu. Soc. Amer.*, **49**, 82–91.

Strickler, P. (1960). "Rückversicherung des Kumulrisikos in der Lebensversicherung." *C.R. XVI Cong. Intern. Actu.*, Brussels, **1**, 666–677.

Taylor, R. H. (1952). "The probability distribution of life annuity reserves and its application to a pension system." *Proc. Confer. Actu. Pub. Practice*, **2**, 100–150.

—— (1967). "A theory of the composite life insurance risk." *Proc. Confer. Actu. Pub. Practice*, **16**, 45–89.

Thyrion, P. (1959). "Sur une propriêté des processus de Poisson généralisées." *Bull. Ass. Roy. Actu. Belges*, **59**, 35–46.

—— (1960). "Note sur les distributions 'par grappes'." *Bull. Ass. R. Actu. Belges*, **60**, 49–66.

—— (1961). "Contribution à l'étude du bonus pour non sinistre en assurance automobile." *Astin Bull.*, **1**, 142–162.

—— (1964). "Les lois exponentielles composées." *Bull. Ass. R. Actu. Belges*, **62**, 35–44.

Tröbliger, A. (1961). "Mathematische Untersuchungen zur Beitragrückgewähr in der Kraftfahrversicherung." *Bl. Deuts. Gesell. Versich. Math.*, **5**, 327–348.

Tucker, H. G. (1963), "An estimate of the compounding distribution of a compound Poisson distribution." *Theory Prob. Appl.*, **8**, 195–200.

Valerius, N. M. (1942). "Risk distributions underlying insurance charges in the retrospective rating plan." *Proc. Casualty Actu. Soc.*, **29**, 96–121.

Walsh, J. E. (1955). "Approximate probability values for observed number of 'successes' from statistically independent binomial events with unequal probabilities." *Sankhyā*, **15**, 281–290.

Welten, C. P. (1963). "Estimation of stop loss premium in fire insurance." *Astin Bull.*, **2**, 356–361.

Wetzel, J. M. (1963). "D'une loi générale de maintien en incapacité de travail." *Bull. Trim. Inst.Actu. Franç.*, **72**, 225–270.

Widder, D. V. (1961). *Advanced Calculus*. Prentice-Hall, Englewood Cliffs, N.J.

Williams, K. (1948). "The classical theory of risk—a statistical approach." *J. Inst. Actu. Stud. Soc.*, **7**, 126–143.

Wold, H. O A. (1937). "A technical study on reinsurance." *C.R. XI Cong. Intern. Actu.*, Paris, **1**, 549–559.

CHAPTER 3

Calculation of "Fair" Net Premiums

PARAMETER ESTIMATION

If the annual claim outgo of a risk business is random variable $X = \sum_{i=1}^{N} X_i$, it would seem fair and reasonable to charge a net or pure premium of $\mathcal{E}X_i$ on the ith contract, assuming that the contract—or its replacement by a contract on an identical risk—will be in force "forever." This is, of course, because the "long run" outgo on this contract averages out to $\mathcal{E}X_i$ per annum and thus the risk business and the contractholder come out even. In particular, if the sampling model deemed appropriate is that discussed at the end of Chapter 2, the fair net premium is λp_1.

On the other hand, the mean $\mathcal{E}X_i$ is no longer the obvious premium if long run considerations do not apply. A prospective contractholder may well be prepared to pay more than $\mathcal{E}X_i$ on a few occasions if there is a possibility that his claim may be many times greater. The risk business would be foolish not to charge this larger premium if it could get it, but competition with other risk businesses may reduce the margin in excess of $\mathcal{E}X_i$ that it is practicable to collect. The subject of conflicting "utilities" of competitive risk business is dealt with in Chapter 6.

In Chapter 2 we considered some of the forms that $F(\cdot)$, the d.f. of X, might assume in practice. If α represents the vector of parameters involved in a particular form, we may write

$$\mathcal{F}\{X \leq x\} = F(x \mid \alpha),$$

and the required net premium is then

$$\kappa_1 \equiv \mathcal{E}X = \int_0^\infty x \, dF(x \mid \alpha). \tag{3.1}$$

An immediate problem is whether we should estimate κ_1 directly or whether there might be advantages in estimating the vector of parameters α.

49

The obvious-estimator of κ_1 is

$$\hat{\kappa}_1 = \frac{1}{n} \sum_{i=1}^{n} x^{(i)}, \tag{3.2}$$

where $x^{(i)}$ is the aggregate outgo of accounting period i ($i = 1, 2, 3, \ldots, n$); but, if, for example, $F(\cdot)$ were Pareto, namely of the form (2.32), with $\alpha \leq 2$ the standard error of $\hat{\kappa}_1$ thus calculated would be infinite and another estimate should be sought.

The problem becomes acute when we wish to calculate the net premium for claims in excess of a fixed amount c; say,

$$\rho_c \equiv \mathfrak{I}\{X > c\}\kappa_{1c} = \int_c^\infty (x - c)\, dF(x \mid \alpha), \tag{3.3}$$

where κ_{1c} is the mean of the truncated distribution of X. An unbiased estimate of κ_{1c} is, by definition, the sample mean of max $(0, X - c)$, where the denominator is the number of positive observations, but if $c \gg \mathcal{E}X$, zero values of $X - c$ will occur frequently and the standard error of this estimate will be large (Vajda, 1951). An alternative is to use the n values of X to estimate the parameters α in $F(x \mid \alpha)$ and to write

$$\hat{\rho}_c = \int_c^\infty (x - c)\, dF(x \mid \hat{\alpha}). \tag{3.4}$$

However, Conolly (1955) has shown that when

$$F'(x \mid \alpha) = \frac{1}{\alpha} e^{-x/\alpha}, \qquad 0 < x < \infty, \tag{3.5}$$

and the mean of n X-values is used to produce an unbiased estimate of α (i.e., $\mathcal{E}\hat{\alpha} = \alpha$), the estimate $\hat{\rho}_c$ derived from (3.4) is biased (unless $c/\alpha = 0$). In fact, in this case

$$\mathcal{E}\hat{\rho}_c > \rho_c,$$

but $\hat{\rho}_c \to \rho_c$ as $n \to \infty$ so that (3.4) is consistent.

The sensitivity of ρ_c to small changes in the vector α led Dubois (1937) to a pessimistic view of the practicability of quoting premiums for so-called stop-loss insurance of aggregate claims in excess of a "priority" c. His viewpoint is still widespread.

An impression of the magnitude of the parameters involved when c is relatively much larger than $\mathcal{E}X$ is obtained by assuming that $F(\cdot)$ is Normal in distribution with mean $\kappa_1 = 0$ and variance $\kappa_2 = 1$ (cp. Vajda, 1955). Choose $c = 2$; then

$$\rho_c = \int_2^\infty (x - 2) \frac{1}{\sqrt{2\pi}} e^{-x^2/2}\, dx = \int_4^\infty \frac{1}{2\sqrt{2\pi}} e^{-z/2}\, dz - 2[1 - \Phi(2)]$$

$$= \Phi'(2) - 2 + 2\,\Phi(2) = 0.0539910 - 2 + 1.9544998 = 0.008491.$$

Let us now simulate a 25-year observational record of values of max $(0, X - 2)$.* Draw two 3-digit random numbers u_1 and u_2, say, and find v_1 and v_2, respectively, from

$$10^{-3}u_1 = \frac{5}{\sqrt{2\pi}} \int_{-\infty}^{v_1} e^{-25t^2/2}\, dt$$

and

$$10^{-3}u_2 = 2^{-12}[\Gamma(12)]^{-1} \int_0^{v_2} e^{-t/2}t^{11}\, dt.$$

Then v_1 is a value of a random variable normally distributed about zero with variance equal to $\frac{1}{25}$ and v_2 is a variate from a chi-square distribution with 24 degrees of freedom. In other words, v_1 is an observed mean based on 25 (years') values and v_2 is an observed sum of 25 squares about an observed mean of 25 normally distributed variates, these two observations being independent of one another.

Suppose that after 25 years' observations of its normally distributed X the risk business decided to reinsure aggregate claims in excess of an estimated priority of the mean plus twice the standard deviation. The point on a $N(0, 1)$ curve corresponding to this priority would be

$$\hat{c} = \frac{2 - v_1}{(v_2/24)^{\frac{1}{2}}}$$

and the corresponding estimate of the net reinsurance premium would be

$$\hat{\rho}_c = \Phi'(\hat{c}) - \hat{c}[1 - \Phi(\hat{c})].$$

As a numerical illustration of this procedure

$$u_1 = 815, \qquad u_2 = 255, \qquad v_1 = 0.179, \qquad v_2 = 28.0,$$
$$\hat{c} = 1.69 \quad \text{and} \quad \hat{\rho}_c = 0.0208.$$

The foregoing simulation of 25 years' observations was carried out 20 times and the results were

mean $v_1 = -0.0522$,	true mean $v_1 = 0$,
mean $v_2 = 23.96$,	true mean $v_2 = 24$,
mean $\hat{c} = 2.204$,	true c-value $= 2$,
mean $\hat{\rho}_c = 0.01072$,	true ρ_c-value $= 0.00849$.
variance $\hat{\rho}_c = 0.00011426$	
$= (0.01069)^2$,	

The estimate of the mean of ρ_c and the estimate of the standard deviation of the distribution of $\hat{\rho}_c$ based on 25 years' results are thus the same. Even 500

---

* William DuMouchel suggested this simple demonstration and carried it out.

years' experience of X-values has produced an estimate of ρ_c that is 26% in excess of the truth.

Nevertheless it is of interest to investigate the possibilities of estimating ρ_c or κ_{1c} for appropriately chosen $c > \mathcal{E}X$. One approach is to apply Bayesian ideas to an assumed parametric form for $F(x \mid \alpha)$; for example, Fürst (1965) assumed that the d.f. of X values in excess of c was

$$F_c(x \mid \alpha) = 1 - e^{(x-c)\theta}, \qquad c < x < \infty, \qquad \theta = \alpha^{-1},$$

and that the prior probability density of θ was given by the gamma distribution

$$\frac{a^m}{\Gamma(m)} \theta^{m-1} e^{-a\theta}. \tag{3.6}$$

Then by Bayes's theorem the posterior density of θ is proportional to

$$\theta^{m+n-1} \exp\left\{-\left[a + \sum_{i=1}^{n} (x_i - c)\right]\theta\right\}, \tag{3.7}$$

given that x_i $(i = 1, 2, \ldots, n)$ is a sample of max $(0, X - c)$ drawn from the distribution $F(x \mid \alpha)$ whose form may be unknown. Before the observations are made, the expected value of θ, which we use as its estimator, is m/a; after the n observations it becomes $(m + n)/[a + \sum_{i=1}^{n} (x_i - c)]$ and the well-known increase in the effect of the observations on the posterior estimate as n increases is manifest.

Two other papers (Hooge, 1965; Jung, 1965) show the effect of using only the $k < n$ (where k may be unity) largest observations of the form max $(0, X - c)$ in estimating the net premium ρ_c. Both authors use asymptotic extreme-value theory (Gumbel, 1958) to avoid assumptions about the form of $F_c(x \mid \alpha)$ or $F(x \mid \alpha)$. It is difficult to think of a practical application of these investigations.

Finally, we refer to an investigation of the values assumed by

$$\rho_{st} = \int_{st}^{\infty} (y - st)\, d_y F(y, t) = \int_{st}^{\infty} [1 - F(y, t)]\, dy,$$

where $F(y, t)$ is given by (2.10) and $p_n(t)$ by (2.28), in the case in which t is large (Esscher, 1965). Following Lundberg (1940; cited in Chapter 2) it can be shown that if

$$\sigma_\rho^2 = \int_{st}^{\infty} (y - st)^2\, d_y F(y, t) - \rho_{st}^2$$

then both

$$\lim_{t \to \infty} \rho_{st} \equiv q(s) \quad \text{and} \quad \lim_{t \to \infty} \frac{\sigma_\rho^2}{t^2} \equiv \sigma_q^2$$

are functions of $U(s)$ and not of $P(\cdot)$. Numerical values of $q(s)$ were calculated by Esscher for six alternative forms of $U(\cdot)$ and were found to differ only slightly relative to σ_q. Esscher also investigated the error in assuming the limit to have been reached for t-values of the order of a few thousands.

CLASSIFICATION OF RISKS

It is well known that the random variable formed by summing N random variables each distributed as Poisson with mean $\lambda_j (j = 1, 2, \ldots, N)$ is itself Poisson with mean $\lambda = \sum_{j=1}^{n} \lambda_j$. If this theorem corresponds to the situation in a portfolio of N contracts, an "average" net premium of λ/N charged to each of the contractholders would provide the correct aggregate amount λ required by the risk business (Franckx, 1964a). In fact a procedure of this type is common in compulsory social insurance programs.

However, if $\lambda_1 < \lambda_2 < \cdots < \lambda_N$ and there were some method of assigning a given contract its place in this series, it would obviously be advantageous to a competitor risk business to provide coverage for the earlier contracts in the series for a premium less than λ/N. The original risk business would then be charging too small a net premium for the remaining contracts. It must therefore protect itself against this possibility by making estimates of the claim probabilities of each (supposedly homogeneous) category of risks classified according to the values assumed by a number of qualitative (e.g., sex, geographic location) or quantitative (e.g., age, milage driven) variables— called "factors" by Almer (1957, cited in Chapter 2). We write p_i for the observed ratio of claims to contractholders in a given year, in which i indexes the risk categories ($i = 1, 2, \ldots, N$), π_i for the corresponding probability of claiming and $\hat{\pi}_i$ for the estimate derived from one of the models discussed later. [If several claims a year occur, quite frequently p_i may exceed unity and the "binomial theory" discussed below may need modification. The reference to Solomon (1948) indicates one possibility.] We suppose that there are q variables which assume the values z_{ji} ($j = 1, 2, \ldots, q$) for every risk in the ith risk classification $i = 1, 2, \ldots, N$; for example, if $j = 2$ represented a classification according to the age of the contractholder, z_{2i} would be the (uniform) age of the contractholders in class i. If $j = 3$ and 4 were used for a qualitative classification into three different geographic locations A, B, and C, z_{3i} would be -1 when class i related to location A, would be $+1$ for location B and zero for location C. Similarly, z_{4i} would assume values -1, 0, $+1$ for locations A, B, and C. A Gaussian linear model to estimate π_i would then be

$$\pi_i = \beta_1 + \sum_{j=2}^{q}(z_{ji} - \bar{z}_j)\beta_j + e_i, \tag{3.8}$$

where $\bar{z}_j = N^{-1} \sum_{i=1}^{N} w_i z_{ji}$, e_i is the value assumed by a random variable E, and w_i is a weight to be specified below. Since p_i may be assumed to be a variate with a binomial distribution having parameters n_i and π_i, where n_i is the number of contracts "exposed" in the ith category, E is approximately Normal about a mean of zero with variance $\pi_i(1 - \pi_i)/n_i$. However, if some of the π_i or n_i are small, this Normality may be far from the truth and it is preferable to base the linear model on the so-called "logit" of π_i so that

$$\tfrac{1}{2} \ln \left(\frac{\pi_i}{1 - \pi_i} \right) = \beta_1 + \sum_{j=2}^{q} (z_{ji} - \bar{z}_j)\beta_j + e_i, \tag{3.9}$$

where E is now Normal with mean zero and variance $\{4n_i\pi_i(1 - \pi_i)\}^{-1} = w_i$, say.

The rationale and fitting of this type of model is described in detail by Draper and Smith (1966) or Seal (1964). Briefly, the q parameters β_j ($j = 1, 2, \ldots, q$) can be estimated as the solution of the q simultaneous linear equations obtained by minimizing the expression

$$\sum_{i=1}^{N} w_i \left[p_i - \beta_1 - \sum_{j=2}^{q} (z_{ji} - \bar{z}_j)\beta_j \right]^2 \tag{3.10}$$

for variation in the β's. The solution will be unique if none of the N-component vectors with typical component $w_i^{1/2} z_{ji}$ for given j can be written as a linear compound of some or all the remaining $q - 1$ such vectors. An immediate difficulty in solving these equations is that the weights w_i involve π_i which is unknown. The correct solution can be achieved iteratively by replacing π_i with p_i and calculating the first set of estimates $\hat{\pi}_i$; then by re-solving with $\hat{\pi}_i$ used in the weight w_i, calculate a second set of estimates $\hat{\hat{\pi}}_i$, and so on. A numerical example in Seal (1968) suggests that four or five iterations will prove adequate. It also illustrates how "interactions" between two z-vectors—namely, to allow for the possibility that, for example, the sex differential in π_i at location A is different from that at location B—may be represented by the product of the corresponding components.

A great advantage of the foregoing model is that individual β's or groups of β can be set equal to zero (or any other chosen values) and the resulting "reduced" model tested for adequacy against the original. In this way, for example, six "interaction" terms were eliminated by Seal in the illustration referred to above.

Perhaps the earliest actuarial example of this kind of analysis is that of Vajda (1943) who applied (3.8) to mortality rates during 1924–1929 among lives insured by British offices. For each of two age groups (separately) there was a three-way classification that consisted of two policy types (whole life and endowment assurance), two dividend groups ("with" and "without"), and six years of observation. All first-order interactions were allowed for in the model. However, n_i was used for w_i instead of the $n_i\pi_i(1 - \pi_i)$ required by binomial theory. In a further communication (1946) that author applied a

2^4 factorial model (namely, one involving four factors each at two "levels") with all interactions up to the third order, the attributes considered being policy type (as before), dividend group (as before), medical examination ("with" and "without"), and policy duration (three and four years versus five years and more). The same author (1947) also applied a linear model in the form of (3.8) to a three-way classification of accidents to a certain type of merchant vessel during World War II. Here there were 22 age groups (treated qualitatively), seven builders, and two "orders" (namely, "first" and "later" casualties), and all first order interactions were again allowed for.* This work was followed by that of Solomon (1948) who avoided the need to iterate to obtain the correct weights by using $\sqrt{\pi_i}$ on the left-hand side of the linear model which results in an approximate weight of $4n_i$. This author fitted cubic polynomials to several sets of data classified in nine ages (46–55) and found that the second- and third-order terms were unnecessary. He also considered the estimation of a factor K, independent of age, which was hypothesized as multiplying the usual binomial variance.

More recently Almer (1957, cited in Chapter 2) proposed the use of a multiplicative factor model first used by Fisher and Mackenzie (1923). The appropriate asymptotic test theory was provided by Philipson (1959) and numerical applications are found in Bailey and Simon (1960), and Mehring (1965). We are doubtful whether there is anything to be gained by this model in comparison with the theoretically simpler Gaussian model which uses the logarithms of the observations.

In fire insurance of property the random variable Y in (2.10) is generally treated as a ratio which represents the "degree of damage" done to the insured property. If s is the sum for which the property is insured, the actual claim paid will then be sy, where y is the value assumed by Y. An interesting feature of this type of insurance is that $P(\cdot)$ can sometimes be written approximately

$$P(y) = \begin{cases} 0, & 0 \le y < \zeta, \\ 1, & \zeta \le y < 1, \end{cases} \tag{3.11}$$

where ζ is some constant. Since $p_1(1)$ is so small, the probability of two or more fires within a year may well be neglected, and it has been found that $p_1(1) p_1$ (in the notation of Chapter 2) may be assumed linear in s. Thus

$$p_1(1) \cdot p_1 \equiv p_s = \alpha + \beta(s - \bar{s}), \tag{3.12}$$

where $\bar{s}$ is the mean value of s in the whole portfolio of properties.

---

* The analysis of variance of a four-way (garage location, usage category, vehicle hp, vehicle model) unbalanced classification of French automobile accidents by Delaporte (1962) is vitiated by (a) its implicit assumption that $n_i \pi_i$ has a variance independent of n_i and π_i and (b) the utilization of an analysis appropriate only for the balanced case.

This model was proposed by Benktander (1953) who fitted (3.12) by least squares to Swedish data classifying each of the following types of property into nine (equidistant?) value classes: (a) wooden barns with hard roofs, (b) villas and cottages, and (c) farm houses. The appropriate chi-square for testing the fit is

$$\sum_s \frac{(y_s - N_s p_s)^2}{N_s p_s (p_2/p_1)}$$

with seven degrees of freedom, where y_s is the aggregate loss in class s and N_s is the number of insured properties in that class. The three values obtained were 8.45, 3.80, and 3.71, which indicate very satisfactory agreement with the model.

EXPERIENCE RATING OF A WHOLE PORTFOLIO BY PAYMENT OF DIVIDENDS

Let us suppose that a risk-loaded premium π_1 is charged for a portfolio of risks with aggregate claim variable X subject to a d.f. $F(\cdot)$ and mean (i.e., net premium) κ_1. Write

$$\pi_1 = \kappa_1 + \eta_1 \equiv \kappa_1(1 + \eta), \tag{3.13}$$

where κ_1 is supposed to be known and invariant and η_1 is the aggregate risk loading. We assume that any expense charges that may be added to π_1 and passed on to the contractholder are exactly equal to the actual expenses of administering the portfolio. The calculation of η_1 is considered in Chapter 5, but in the meantime we suppose that it is more than adequate in the long run and that consideration is being given to returning a dividend to the contractholders based on the actual claims of the portfolio. Note that there is no question in our minds that we have incorrectly estimated κ_1. It is only the continuation of surpluses arising from the size of η_1 that is prompting the payment of dividends. This, of course, immediately suggests that account should be taken of the size of the risk reserve in determining the dividend. Such an approach is, indeed, found in Chapter 6, but in the meantime we proceed by means of long-run expectations.

In the following "over-all" technique we do not consider the allocation of the aggregate dividend among the individual contractholders: their possible subdivision into "good boys" and "bad boys" is considered below. For the time being we may regard the portfolio as a one-year "group" insurance paid for by a single contractholder The case in which the surviving members of a group of identical long-term contracts are to receive gradual repayment of risk loadings rendered unnecessary by favorable emerging experience has been treated in an early paper by Vajda (1930).

Consider, first, the expected surplus before payment of a dividend. It is, by definition,

$$\int_0^{\pi} (\pi_1 - x)\, dF(x) = \int_0^{\pi_1} dF(x) \int_x^{\pi_1} dz = \int_0^{\pi_1} dz \int_0^z dF(x) = \int_0^{\pi_1} F(z)\, dz.$$

Similarly, the expected loss is

$$\int_{\pi_1}^{\infty} (x - \pi_1)\, dF(x) = \int_{\pi_1}^{\infty} [1 - F(z)]\, dz$$

and the expected net profit (before dividend payment) is

$$\int_0^{\pi_1} F(z)\, dz - \int_{\pi_1}^{\infty} [1 - F(z)]\, dz = \int_0^{\pi_1} dz - \int_0^{\infty} [1 - F(z)]\, dz = \pi_1 - \kappa_1 = \eta_1,$$

as it should be.

Ammeter (1961) has suggested that a general type of dividend formula would provide dividends equal to the values assumed by the random variable

$$D_+ = \max(0, a\pi_1 - bX), \tag{3.14}$$

where no restriction is set on a and b other than that they are to be positive. Then

$$\mathcal{E}D_+ = \int_0^{a\pi_1/b} (a\pi_1 - bx)\, dF(x) = b \int_0^{a\pi_1/b} F(x)\, dx \tag{3.15}$$

and

$$\mathcal{V}D_+ = \int_0^{a\pi_1/b} (a\pi_1 - bx)^2\, dF(x) - (\mathcal{E}D_+)^2$$

$$= 2b^2 \int_0^{a\pi_1/b} \left(\frac{a\pi_1}{b} - x\right) F(x)\, dx - b^2 \left[\int_0^{a\pi_1/b} F(x)\, dx\right]^2. \tag{3.16}$$

It is convenient also to introduce

$$D_- = \min(0, a\pi_1 - bX),$$

so that

$$D_+ - D_- = a\pi_1 - bX,$$

$$\mathcal{E}D_+ - \mathcal{E}D_- = a\pi_1 - b\kappa_1,$$

$$\mathcal{V}(D_+ - D_-) = \mathcal{V}D_+ + \mathcal{V}D_- - 2\mathcal{C}D_+D_- = b^2\mathcal{V}X$$

and, since

$$\mathcal{C}D_+D_- = \mathcal{E}D_+D_- - (\mathcal{E}D_+)(\mathcal{E}D_-) = -\mathcal{E}D_+(\mathcal{E}D_+ - a\pi_1 + b\kappa_1)$$

$$= (a\pi_1 - b\kappa_1)\mathcal{E}D_+ - (\mathcal{E}D_+)^2,$$

$$\mathcal{V}D_- = b^2\mathcal{V}X - \mathcal{V}D_+ + 2(a\pi_1 - b\kappa_1)\mathcal{E}D_+ - 2(\mathcal{E}D_+)^2. \tag{3.17}$$

A "natural" dividend system would then be one under which the expected dividend would equal the premium margin so that

$$\mathcal{E}D_+ = \pi_1 - \kappa_1 = \eta_1. \tag{3.18}$$

In general, this relation provides a single equation for a and b, but, as we shall see, when $F(\cdot)$ ia a Normal d.f. both parameters can be determined. Jackson (1953) proposed this relation with $a = b$, and Ammeter (1957) with $b = 1$. The disadvantage of this "natural" system, which returns the whole risk loading, is that the risk reserve will eventually become negative (see Chapter 4). In lieu of (3.18) we might therefore prefer to utilize Jackson's (1953) alternative equation (with $a = b$)

$$\mathcal{E}D_{++} = \int_0^{\pi_1} (\pi_1 - x)\, dF(x) - \int_{\pi_1'}^{\infty} (x - \pi_1')\, dF(x)$$

$$= \pi_1' - \kappa_1 + \int_{\pi_1'}^{\pi_1} F(x)\, dx, \qquad (3.19)$$

where $\pi_1' < \pi_1$. The expected surplus has here been offset by losses in excess of a smaller premium than that actually received. As another alternative, Jackson suggested writing $b = 1$ on the left-hand side of (3.18) and replacing $a\pi_1$ with $\pi_1 - c$.

Ammeter (1957) suggested the use of π_1' in lieu of π_1 in (3.18). The expected contribution to the risk reserve is then $\pi_1 - \pi_1'$ instead of $\pi_1 - \kappa_1 = \eta_1$. As a third or "conservative" dividend system, Ammeter suggested the relation

$$\mathcal{E}D_+ + c\sqrt{\mathcal{V}D_-} = \pi_1 - \kappa_1, \qquad (3.20)$$

where c might be chosen as unity. Here a multiple of the standard deviation of the losses reduces the premium margin available for dividend distribution.

In general $\mathcal{E}D_+$ can only be obtained numerically by approximate methods, but the calculations are relatively simple when $F(x)$ can be assumed to be Normal. Suppose, for example, that the mean and variance of this distribution are given by (2.41) with $\lambda = 1$ for convenience and the mean individual claim is equal to unity (i.e., $p_1 = 1$). Then

$$\kappa_1 = t, \qquad \kappa_2 = tp_2, \qquad \pi_1 = t + \eta_1 \equiv (1 + \eta)t,$$

and we write bd for the standardized value of the dividend when claims equal their expectation; namely

$$bd = \frac{a(1 + \eta)t - bt}{\sqrt{tp_2}}. \qquad (3.21)$$

Then

$$\mathcal{E}D_+ = \int_0^{a\pi_1/b} (a\pi_1 - bx)\, dF(x) = b\int_0^{t + d\sqrt{tp_2}} (t + d\sqrt{tp_2} - x)\, dF(x)$$

$$= b\int_{-\infty}^{t + d\sqrt{tp_2}} (t + d\sqrt{tp_2} - x)\frac{1}{\sqrt{2\pi tp_2}} \exp\left[\frac{-(x - t)^2}{2tp_2}\right] dx$$

[Ammeter (1961) prefers to ascribe a constant dividend of $b(t + d\sqrt{tp_2})$ to all $x < 0$]

$$= \frac{b\sqrt{tp_2}}{\sqrt{2\pi}} \int_{-\infty}^{d} (d - z) \exp\left(\frac{-z^2}{2}\right) dz$$

$$= b\sqrt{tp_2}\, [d\, \Phi(d) + \Phi'(d)]. \qquad (3.22)$$

Further

$$\mathcal{V}D_+ = \int_0^{a\pi_1/b} (a\pi_1 - bx)^2 \, dF(x) - (\mathcal{E}D_+)^2,$$

where

$$\int_0^{a\pi_1/b} \left(\frac{a\pi_1}{b} - x\right)^2 dF(x) = tp_2 \int_{-\infty}^d (d-z)^2 \frac{1}{\sqrt{2\pi}} \exp\left(\frac{-z^2}{2}\right) dz$$

$$= tp_2[(1 + d^2)\,\Phi(d) + d\,\Phi'(d)], \qquad (3.23)$$

since

$$\frac{1}{\sqrt{2\pi}} \int_{-\infty}^d z^2 e^{-z^2/2} \, dz = \int_{-\infty}^d z^2 \, d\Phi(z) = -\int_{-\infty}^d z \, d\Phi'(z),$$

$$\text{because } \Phi''(z) = -z\,\Phi'(z),$$

$$= -z\,\Phi'(z) \Big|_{-\infty}^d + \int_{-\infty}^d \Phi'(z) \, dz$$

$$= -d\,\Phi'(d) + \Phi(d).$$

As a numerical illustration, we take $\kappa_2 = 2t$, which would be appropriate if the distribution of individual claims were negative exponential, for example, and assume that (except for the "conservative" system) three-quarters of a 20% net premium loading is to be distributed in dividends. We note that

$$\int_{\pi_1'}^{\pi_1} F(x) \, dx = \int_{1.15t}^{1.2t} \Phi\left\{\frac{x-t}{\sqrt{2t}}\right\} dx = \int_{0.15\sqrt{t/2}}^{0.2\sqrt{t/2}} \Phi(z) \, dz$$

$$= z\,\Phi(z) \Big|_{0.15\sqrt{t/2}}^{0.2\sqrt{t/2}} - \int_{0.005625t}^{0.01t} e^{-x} \, dx \equiv A(t),$$

$$\mathcal{E}D_+ = \sqrt{2t}\, b[d\,\Phi(d) + \Phi'(d)] \equiv \sqrt{2t}\, b f_1(d),$$

$$\mathcal{V}D_+ = 2tb^2\{(1 + d^2)\,\Phi(d) + d\,\Phi'(d) - [d\,\Phi(d) + \Phi'(d)]^2\}$$

$$\equiv 2tb^2 f_2(d),$$

$$\mathcal{V}D_- = 2tb^2\{(1 + d^2)[1 - \Phi(d)] - d\,\Phi'(d) - [d(1 - \Phi(d)] - \Phi'(d)^2\}$$

$$= 2tb^2 f_2(-d)$$

and

$$d = \left(\frac{1.2a}{b} - 1\right)\sqrt{\frac{t}{2}}.$$

Jackson's relation (3.19) then becomes

$$\sqrt{2t}\, b f_1\!\left(\frac{\cdot 2\sqrt{t}}{2}\right) = 0.15t + A(t), \qquad \text{with} \quad a = b. \tag{3.19}'$$

Ammeter's version of (3.18) is

$$\sqrt{2t}\, b f_1(d) = 0.15t, \qquad \text{with } d \text{ as above.} \tag{3.18}'$$

As alternatives in the equation last written we can (a) replace d with $0.2\sqrt{t/2}$ (i.e., $a = b$) and (b) find d and thus a by putting $b = 1$.

Table 3.1

Relation	d	$t = 8$			$t = 2048$		
		a	b	$\dfrac{\sqrt{\mho D_+}}{\mathcal{E}D_+}$	a	b	$\dfrac{\sqrt{\mho D_+}}{\mathcal{E}D_+}$
$(3.19)'$	$0.2\sqrt{t/2}$	0.493	0.493	1.133	0.754	0.754	0.156
$(3.18)'$	$0.2\sqrt{t/2}$	0.476	0.476	1.133	0.750	0.750	0.156
	$(1.2a - 1)\sqrt{t/2}$	0.743	1.000	1.698	0.958	1.000	0.208
	-2.0	$\cdots$	35.332	8.889	441.654	565.318	8.889
	-1.2	1.782	5.347	3.764	68.624	85.558	3.764
	-0.4	0.868	1.302	1.937	14.284	17.358	1.937
	0	0.627	0.752	1.463	10.027	12.032	1.463
	0.4	0.476	0.476	1.133	6.424	7.614	1.133
	1.2	0.318	0.239	0.718	3.304	3.821	0.718
	2.0	0.249	0.149	0.493	2.116	2.390	0.493
$(3.20)'$	-2.0	inapplicable ($a = 0$)			5.010	6.412	8.889
	-1.2	0.139	0.418	3.764	5.355	6.677	3.764
	-0.4	0.282	0.423	1.937	5.575	6.775	1.937
	0	0.339	0.407	1.463	5.427	6.512	1.463
	0.4	0.371	0.371	1.133	5.015	5.943	1.133
	1.2	0.363	0.273	0.718	3.771	4.362	0.718
	2.0	0.320	0.192	0.493	2.719	3.071	0.493

Finally, Ammeter's "conservative" method gives

$$\sqrt{2t}\, b[f_1(d) + \sqrt{f_2(-d)}] = 0.2t, \qquad \text{with } c = 1 \text{ and } d \text{ as above.} \tag{3.20}'$$

Table 3.1 provides some illustrations of the foregoing results. Note that he coefficient of variation of any of the above dividend formulas is

$$\frac{\sqrt{\mho D_+}}{\mathcal{E}D_+} \sim \frac{1}{d} \quad \text{as} \quad d \to \infty \tag{3.24}$$

and decreases toward zero as d increases without bound. We observe that although values of a and b in excess of unity may be difficult to interpret there is nothing illogical about them.

Suppose that π_1' is substituted for π_1 in (3.18) and the Normal approximation used for $F(\cdot)$ then dividend formulas with small coefficients of variation are produced when $d \to \infty$ and $f_1(d) \sim d$, which implies that

$$b \sim \frac{\pi_1' - t}{d\sqrt{tp_2}} = \frac{\eta'}{d}\sqrt{\frac{t}{p_2}}, \quad \text{where} \quad \pi_1' = (1 + \eta')t, \quad \eta' < \eta, \quad (3.25)$$

and

$$a \sim \frac{\eta'}{1 + \eta}\left(1 + \frac{1}{d}\sqrt{\frac{t}{p_2}}\right) \sim \frac{\eta'}{1 + \eta}, \quad \text{unless} \quad t = O(d^2); \quad (3.26)$$

for example, when $d = 10^6$ in the foregoing illustration and $t = 200$,

$$b \sim 0.15 \times 10^{-5} \quad \text{and} \quad a \sim 0.125.$$

A very "smooth" dividend formula with an expected return of three-quarters of the 20% risk loading would thus be

$$D_+ = \max\,(0, 0.125\pi_1 - 0.15 \times 10^{-5}X),$$

since it would tend to return 14.375% of the 15% dividend loading every year unless claims were extraordinarily large.

Although larger (a, b) pairs have the disadvantage of larger coefficients of variation in the dividend payments, it must be remembered that the larger the b-value, the more the weight given to actual experience. These larger b-values thus tend to protect the risk business against errors in the estimation of κ_1, the expected aggregate claim outgo.

The concept that the choice of a suitable pair (a, b) could be made to depend on the behavior of $\mathcal{E}D_+$ under changes in the value of κ_1 was investigated by Ammeter (1961). Retaining the hypothesis of Normality for X, we have to consider

$$\mathcal{E}D_+(t) = b\sqrt{tp_2}\,[d\,\Phi(d) + \Phi'(d)] \quad \text{with} \quad d = \left(\frac{1 + \eta a}{b} - 1\right)\left(\frac{t}{p_2}\right)^{1/2}$$
$$(3.27)$$

as a function of $\kappa_1 = t$ for alternative pairs (a, b) calculated from the relations

$$b\sqrt{t^0 p_2}\,[d^0\,\Phi(d^0) + \Phi'(d^0)] = \eta' t^0 \quad \text{and} \quad d^0 = \left(\frac{1 + \eta a}{b} - 1\right)\left(\frac{t^0}{p_2}\right)^{1/2},$$
$$(3.28)$$

t^0 being the "expected" value of κ_1 on which the dividend system is to be based.

Ammeter calls any pair (a, b) "unbiased" at the point t if

$$\mathcal{E}D_+(t) = (1 + \eta')t^0 - t, \tag{3.29}$$

namely, if the expected dividend equals the actual margin in the premium available for dividend purposes. All pairs (a, b) are unbiased at the point $t = t^0$, and the problem is to find those pairs that are unbiased for as wide a range of t-values as possible.

We note that (3.29) is a straight line on the $[t, \mathcal{E}D_+(t)]$ graph with slope -1 and that the premium margin is negative once $t > (1 + \eta')t^0$. An interesting (a, b)-pair is found by assuming that both t^0 and t are large and d^0 is chosen equal to $\eta'\sqrt{t^0/p_2}$. The relations (3.25) and (3.26) then lead to

$$b \sim 1 \quad \text{and} \quad a \sim \frac{1 + \eta'}{1 + \eta} \tag{3.30}$$

and (3.27) becomes

$$\mathcal{E}D_+(t) \sim \eta't. \tag{3.31}$$

The upward-sloping straight line (3.31) intersects with (3.29) only at $t = t^0$, but since η' is relatively small it remains close to (3.29) for a range of values of t in the neighborhood of t^0. When t is actually smaller than t^0, the expected dividend is smaller than the premium margin, but when $t > t^0$ dividends tend to be larger than planned and might seriously deplete the risk reserve.

Consideration of the curve $\mathcal{E}D_+(t)$ suggests that it is important to decide whether the estimate of t^0 is on the high or the low side. Relation 3.27 gives

$$2\frac{d\mathcal{E}D_+(t)}{dt} = b\sqrt{\frac{p_2}{t}}\,[d\,\Phi(d) + \Phi'(d)]$$

$$+ b\sqrt{tp_2}\,[d\,\Phi'(d) + \Phi(d) + \Phi''(d)]\sqrt{\frac{1}{tp_2}}\left[\frac{1 + \eta \cdot a}{b} - 1\right]$$

$$= b\sqrt{\frac{p_2}{t}}\,[2d\,\Phi(d) + \Phi'(d)],$$

Since

$$\Phi''(d) = -d\,\Phi'(d),$$

which is negative for $d < -0.6120$ and positive otherwise. The curve of $\mathcal{E}D_+(t)$ is thus increasing or decreasing at the point $t = t^0$ according as $d^0 \gtrless -0.6120$.

Therefore, if management of a risk business thinks that its t^0 is on the high rather than the low side—and business prudence suggests that this is likely—it is desirable that $\mathcal{E}D_+(t)$ decrease at $t = t^0$. If we further require that the curve of $\mathcal{E}D_+(t)$ remain unbiased for t-values slightly less than t^0, its angle with the t-axis should be $45°$ when $t = t^0$. Hence

$$b\sqrt{\frac{p_2}{t^0}}\,[2d^0\,\Phi(d^0) + \Phi'(d^0)] = -2,$$

whereas (3.28) requires

$$b\sqrt{t^0 p_2}[d^0\,\Phi(d^0) + \Phi'(d^0)] = \eta' t^0.$$

On dividing the second of these relations by the first

$$\frac{d^0\,\Phi(d^0) + \Phi'(d^0)}{2d^0\,\Phi(d^0) + \Phi'(d^0)} = \frac{-\eta'}{2}. \qquad (3.32)$$

This is an equation which determines d^0 uniquely from η'. Now the negative quantity on the left of this equation decreases as $d^0 < -0.6120$ decreases; for example, with $d^0 = -3.0$ the left-hand side equals -0.104 and produces a dividend loading of $100\,\eta'\% = 20.8\%$, which may be considered fairly large. When d^0 is -3.0 or smaller (i.e., when the dividend loading $\leq 20.8\%$), we may use an asymptotic relation for $\Phi(d^0)$, namely,

$$\Phi(d^0) \sim \frac{\Phi'(d^0)}{-d^0}\left[1 - \frac{1}{2 + (d^0)^2}\right].$$

Relation 3.32 then becomes

$$(d^0)^2 \sim \frac{2}{\eta'}$$

(which, for example, with $d^0 = -3$ produces $\eta' \sim 0.222$ instead of the exact $\eta' = 0.208$), and (3.28) gives the following explicit values for a and b:

$$b \sim 2\sqrt{\frac{t^0}{p}}(1 + \eta')\left[\Phi'\left(\frac{2}{\eta'}\right)^{\frac{1}{2}}\right]^{-1}, \qquad (3.33)$$

$$a \sim 2\frac{1 + \eta'}{1 + \eta}\left[\left(\frac{t^0}{p_2}\right)^{\frac{1}{2}} + \left(\frac{2}{\eta'}\right)^{\frac{1}{2}}\right]\left[\Phi'\left(\frac{2}{\eta'}\right)^{\frac{1}{2}}\right]^{-1};$$

for example, with $\eta = 0.2$, $\eta' = 0.15$, and $p_2 = 2$ (as before), (3.33) results in $a \sim 21{,}335$ and $b \sim 9{,}060$, when $t^0 = 8$, and $a \sim 134{,}565$ and $b \sim 144{,}939$, when $t^0 = 2048$. Although these large a and b values may be theoretically reasonable, the writer is sceptical of their practical utility.

Ammeter (1961) considers (a, b)-pairs that tend to be unbiased on both sides of $t = t^0$ rather than on the positive side only and the reader is referred to the original for his results.

EXPERIENCE RATING THE PREMIUM OF AN INDIVIDUAL CONTRACTHOLDER

Under the "proneness" model the number of claims made by a contractholder during an accounting period of length t is Poisson distributed with mean λt, where λ is a particular value of a random variable Λ which has d.f. $U(\cdot)$. Given N contractholders, we may suppose their λ-values to be a sample of N from $U(\cdot)$ and we may attempt to assign a specified contractholder to his proper value on the scale of Λ by observing the numbers of his claims in one or more accounting periods. Although in Chapter 2 accident data have been used to estimate the parameters of appropriate forms of $U(\cdot)$, we now have the different problem of estimating the λ-value of an individual from his accident record of one or more years and of using this estimate to calculate his new premium rate. Casualty actuaries in the United States refer to this technique as *prospective* experience rating in contradistinction to the *retrospective* experience rating by means of dividend or bonus payments.

Consider, first, how an initial estimate of the parameters of the distribution of "proneness," namely $U(\cdot)$, is changed by the information that there were n claims during a period of length t. By Bayes's theorem

$$dU(\lambda \mid n, t) = \frac{e^{-\lambda t}[(\lambda t)^n / n!] \, dU(\lambda)}{\displaystyle\int_0^\infty e^{-\lambda t}[(\lambda t)^n / n!] \, dU(\lambda)}.$$

$$= e^{-\lambda t} \frac{(\lambda t)^n}{n!} \frac{dU(\lambda)}{p_n(t)} \qquad (3.34)$$

If, for example, it could be assumed that

$$U'(\lambda) = \frac{c^m}{\Gamma(m)} e^{-c\lambda} \lambda^{m-1}, \qquad c, m > 0; \qquad 0 < \lambda < \infty,$$

so that

$$\mathcal{E}\Lambda = \frac{m}{c} \quad \text{and} \quad \mathcal{U}\Lambda = \frac{m}{c^2},$$

and, as shown in Chapter 2,

$$p_n(t) = \binom{m + n - 1}{n} \left(\frac{c}{c + t}\right)^m \left(\frac{t}{c + t}\right)^n, \qquad n = 0, 1, 2, \ldots, \quad (3.35)$$

then

$$U'(\lambda \mid n, t) = \frac{(c + t)^{m+n}}{\Gamma(m + n)} e^{-(c+t)\lambda} \lambda^{m+n-1}. \qquad (3.36)$$

The effect of the observation of the n claims in the period of length t is thus to change the mean and variance of the distribution of Λ to

$$\mathcal{E}(\Lambda \mid n, t) = \frac{m + n}{c + t} \quad \text{and} \quad \mathcal{U}(\Lambda \mid n, t) = \frac{m + n}{(c + t)^2},$$

respectively. As t, and thus n, increase

$$\mathcal{E}(\Lambda \mid n, t) \to \frac{n}{t} \quad \text{and} \quad \mathcal{U}(\Lambda \mid n, t) \to \frac{n}{t^2},$$

thus showing that the distribution of Λ is eventually independent of the initial values set on the parameters m and c. We note that it follows that the distribution of claims in a period of length s following the observation of n claims in a period of length t is (3.35), that is, negative binomial, with c replaced by $c + t$ and m replaced by $m + n$ (Dropkin, 1960).

With the gamma distribution for $U(\cdot)$, the initial value of m/c would be a claim rate based perhaps on the experience of an analogous class of risks, and the variance of the distribution of Λ is then $1/c = b^2$ times this mean claim rate. Keffer (1929) used these ideas to reclassify contracts covering groups of employees for one-year-term life insurance. The expected claim rate for any contract was determined by aggregating the individual values of $s_i q_i$, where s_i is the sum insured on individual i and q_i is his probability of death within a year based on a standard table. The discussion of the paper (*op. cit.*, pp. 593–611) is interesting because of the wide range of estimates of c that were proposed. J. B. Glenn reported that experience with 61 groups in the mineral oil industry gave $c \simeq 9$, whereas the author's experience suggested a concentration of c-values between 4 and 24. Similar considerations were applied independently by Lundberg (1941) to individual sickness indemnity policies. He recommended that t be fixed at about five years to avoid unnecessary fluctuations in the premium rate.

Reverting to the general case, we find that the probability that an individual (contract) with invariant "proneness" will make l claims during a period of length s following on a period of length t during which he made n claims is

$$p_{l|n}(s \mid t) = \frac{\int_0^\infty e^{-\lambda(t+s)} \dfrac{(\lambda t)^n (\lambda s)^l}{n! \, l!} \, dU(\lambda)}{\int_0^\infty e^{-\lambda t} \dfrac{(\lambda t)^n}{n!} \, dU(\lambda)}. \tag{3.37}$$

The expected value of L under these circumstances is thus

$$\mathcal{E}(L, s \mid n, t) = \sum_{l=0}^\infty l \, p_{l|n}(s \mid t) = \int_0^\infty e^{-\lambda(t+s)} \frac{(\lambda t)^n}{n!} \, dU(\lambda) \sum_{l=0}^\infty \frac{l(\lambda s)^l / l!}{p_n(t)}$$

$$= \frac{s}{t}(n+1) \frac{p_{n+1}(t)}{p_n(t)} \tag{3.38}$$

This is an increasing function of n and a decreasing function of t and determines $U(\cdot)$ uniquely. It is only linear in n if $U(\cdot)$ is a gamma distribution (Lundberg, 1940, cited in Chapter 2; Johnson, 1957, 1967).

A further result is that by (3.34)

$$\mathcal{E}(\Lambda \mid n, t) = \int_0^\infty \lambda e^{-\lambda t} \frac{(\lambda t)}{n!} \frac{dU(\lambda)}{p_n(t)} = \frac{n+1}{t} \frac{p_{n+1}(t)}{p_n(t)}$$

$$= \frac{1}{s} \mathcal{E}(L, s \mid n, t). \tag{3.39}$$

Lundberg calls this the "intensity function." When $t \to \infty$, $\mathcal{E}(L, s \mid n, t)$ thus tends to s times the "true proneness" of the individual contract under observation. If $U(\cdot)$ is a gamma distribution,

$$\frac{\mathcal{E}(\Lambda \mid n, t)}{\mathcal{E}\Lambda} = \frac{1 + n/m}{1 + t/c} \equiv M(n, t), \qquad (3.40)$$

and a table of this function for $n = 0, 1, 2, \ldots$, and $t = 1, 2, 3, \ldots$, shows how the original premium is modified by accumulating experience (Bichsel, 1964). In particular, if $n = 0$, the multiplier of the original premium is $(1 + t/c)^{-1}$ (Hewitt, 1960; Bichsel, 1961).

Delaporte (1965) calls $\mathcal{E}(L, 1 \mid n, t)$ "la prime modelée sur le risque" and has provided a number of illustrations of the foregoing theory based on French automobile accident data and the assumption that $U(\cdot)$ is a gamma distribution. He has proposed a measure of the efficiency of the above posterior procedure in comparison with the use of $\mathcal{E}\Lambda$, but Lundberg (1966) considers that a better index is the ratio of the mean conditional variance of L at time t, the weights being $p_n(t)$, to the initial variance before observations are made.

In order to apply the posterior procedure in practice, Delaporte (1965) proposed the following:

1. The contract should contain a table of multipliers $M(n, t)$ from (3.40).

2. Because of the danger that large M-values will result in cancellation of the contract by the insured, they should be modified to $M'(n, t)$ in such a way that the mean premium is unchanged, namely,

$$\sum_{n=0}^{\infty} M(n, t)\, p_n(t) = \sum_{n=0}^{\infty} M'(n, t)\, p_n(t)\, p(M') \geq \sum_{n=0}^{\infty} M(n, t)\, p_n(t)\, p(M),$$

where $p(M)$ is the probability that the premium will be paid if the multiplier assumes the value M; for example, if experience has revealed a cancellation threshold at one-and-a-half times the standard premium

$$p(M) = \begin{cases} 1, & M < 1.5, \\ 0, & M \geq 1.5. \end{cases}$$

3. In order to have a competitive first year's premium, it should be set at, say, $0.85\mathcal{E}\Lambda$ instead of $\mathcal{E}\Lambda$ and the 15% loss should be recouped over the next four years with due allowance made for cancellation of contracts even in cases in which M is less than the cancellation threshold.

Instead of aggregating the numbers of claims made during a t-year period, we may base the estimate of a contractholder's λ-value on his current "dividend class" $z(t)$ which depends only on $\{i_\tau\}$, the set of claim-free-year indicators $i_\tau (\tau = 1, 2, \ldots, t)$, where i_τ is unity if there are no claims in year

τ and zero if one or more claims occur (Pesonen, 1963). We are thus interested in calculating $\mathcal{E}(\Lambda \mid z(t))$ and write $p(z \mid i)$ for the proportionate number of ways of reaching dividend class z with i claim-free years so that

$$\sum_z p(z \mid i) = 1 \quad \text{and} \quad i = \sum_{\tau=1}^{t} i_\tau.$$

We consider only the particular case in which $U(\cdot)$ is a gamma distribution, namely

$$U'(\lambda) = \frac{c^m}{\Gamma(m)} e^{-c\lambda} \lambda^{m-1}, \qquad c, m > 0, \qquad 0 < \lambda < \infty;$$

then

$$U'(\lambda \mid z(t)) = \frac{\displaystyle\sum_{i=0}^{t} p(z \mid i) \binom{t}{i} e^{-\lambda i}(1 - e^{-\lambda})^{t-i} U'(\lambda)}{\displaystyle\sum_{i=0}^{t} p(z \mid i) \binom{t}{i} \int_0^{\infty} e^{-\lambda i}(1 - e^{-\lambda})^{t-i} U'(\lambda)\, d\lambda}, \qquad (3.41)$$

where

$$\int_0^{\infty} e^{-\lambda i}(1 - e^{-\lambda})^{t-i} U'(\lambda)\, d\lambda = \sum_{j=0}^{t-i} \binom{t-i}{j} (-1)^j \frac{c^m}{\Gamma(m)} \int_0^{\infty} e^{-(i+j+c)\lambda} \lambda^{m-1}\, d\lambda$$

$$= \sum_{j=0}^{t-i} \binom{t-i}{j} (-1)^j c^m (i + j + c)^{-m}.$$

Hence

$$\mathcal{E}(\Lambda \mid z(t)) = \frac{\displaystyle\sum_{i=0}^{t} \binom{t}{i} p(z \mid i) \int_0^{\infty} \lambda e^{-\lambda i}(1 - e^{-\lambda})^{t-i} U'(\lambda)\, d\lambda}{\displaystyle\sum_{i=0}^{t} \binom{t}{i} p(z \mid i) \int_0^{\infty} e^{-\lambda i}(1 - e^{-\lambda})^{t-i} U'(\lambda)\, d\lambda}$$

$$= \frac{\displaystyle\sum_{i=0}^{t} \binom{t}{i} p(z \mid i) m \sum_{j=0}^{t-i} \binom{t-i}{j}(-1)^j (i + j + c)^{-m-1}}{\displaystyle\sum_{i=0}^{t} \binom{t}{i} p(z \mid i) \sum_{j=0}^{t-i} \binom{t-i}{j}(-1)^j (i + j + c)^{-m}}. \qquad (3.42)$$

As an example, we consider a dividend classification into five classes. The contractholder starts in the lowest class, class A. He moves one class forward for every no-claim year until he reaches the highest class (with the largest dividend), class E, in which he remains until he makes a claim. If a contractholder is above class B when he has one or more claims in a year, he moves back to class B for his following year's premium calculation. If he is in class B when he has one or more claims, he moves back to class A, in which he remains if he has another year with one or more claims.

Simple calculations provide the following values for $p(z \mid i)$ at the end of five years.

z	i	$p(z \mid i)$
E	5	1
E	4	$\frac{1}{5}$
D		$\frac{2}{5}$
C		$\frac{1}{5}$
B		$\frac{1}{5}$
D	3	$\frac{1}{10}$
C		$\frac{3}{10}$
B		$\frac{5}{10}$
A		$\frac{1}{10}$
C	2	$\frac{1}{10}$
B		$\frac{4}{10}$
A		$\frac{5}{10}$
B	1	$\frac{1}{5}$
A		$\frac{4}{5}$
A	0	1

If accumulating experience does not modify the original parameters which we will assume are $m = 2$ and $c = 1$, so that the mean number of claims per contract (per annum) is $m/c = 2$, the numerator of $\mathcal{E}(\Lambda \mid z = C)$, for example, is twice

$$\tfrac{5}{5}(5^{-3} - 6^{-3}) + \tfrac{30}{10}(4^{-3} - 2 \times 5^{-3} + 6^{-3})$$
$$+ \tfrac{10}{10}(3^{-3} - 3 \times 4^{-3} + 3 \times 5^{-3} - 6^{-3}),$$

and the denominator has the same form except that the index of the power to which the individual terms are raised is -2 instead of -3. The result is

$$\mathcal{E}(\Lambda \mid C) = \frac{2 \times 0.0256667}{0.0588889} = 0.8717,$$

and similar calculations provide

$$\mathcal{E}(\Lambda \mid A) = 2.4579.$$

Compared with an initial net premium of 2 for all contracts, the individual who is still in class A after five years should pay a premium of 2.46, whereas a contractholder in class C should pay only 0.87.

Pesonen (loc. cit.) suggests that a suitable bonus scale is provided by the vector $\lim_{t \to \infty} \mathcal{E}(\Lambda \mid z(t))$ if the limit exists. He also considers a more general form of $U(\cdot)$ which consists of a mixture of gamma distributions with fixed m and shows how $\mathcal{E}(\Lambda \mid z(t))$ is changed when there is an influx of new contractholders at a constant rate of α times the existing portfolio.

If the dividend system depends only on i, the number of claim-free years out of t, the calculations are particularly simple. We see from (3.36) that if every year has been a no claim year, namely, if $i = t$,

$$U(\lambda \mid 0, t) = \int_0^\lambda \frac{(c + t)^m}{\Gamma(m)} e^{-(c+t)z} z^{m-1} \, dz$$

$$\equiv P(c + t\lambda, m) \tag{3.43}$$

in comparison with the initial d.f.

$$U(\lambda) = P(c\lambda, m). \tag{3.44}$$

Table 3.2 provides some comparisons of the 5, 50, and 95% points of these

Table 3.2   Values of λ Calculated from (3.43)

t	$m = 1.2$			$m = 1.5$		
	5%	50%	95%	5%	50%	95%
0	0.169	1.615	6.133	0.141	0.946	3.126
1	0.089	0.846	3.212	0.101	0.676	2.233
2	0.060	0.573	2.176	0.078	0.526	1.737
3	0.045	0.433	1.645	0.064	0.430	1.421
4	0.037	0.348	1.323	0.054	0.364	1.202
5	0.031	0.291	1.106	0.047	0.315	1.042

distributions based on the pairs $c = 1.1$ and $m = 1.2$ and $c = 2.5$ and $m = 1.5$ deduced by Delaporte (1957) from French automobile data.* The progressive decrease in the effective range of λ-values as t increases is noteworthy, but it is the first year's no-claim result that has the most effect. When $m = 1.2$, the median λ-value decreases to less than one-fifth of its initial value in five years, although the reduction is only to one-third of the initial median when $m = 1.5$.

The more general approach of Bühlmann (1964) does not require that λ be a Poisson parameter and can be applied directly to the amounts of claim made in successive contract years. Write $P(y \mid \lambda)$ for the d.f. of Y, the random variable that represents the claim amount under a specified contract; in particular $P(0 \mid \lambda)$ is the probability that no claim will occur in a given contract year. This distribution is assumed to be time-invariant and successive Y-values are supposed to be independently distributed.

Write $R_n^0(Y_1, Y_2, \ldots, Y_n) \equiv R_n^0$ for the optimal estimate of λ based on claim amounts observed in n successive contract years. Let $L(\cdot)$ be a "loss function" of the difference between an estimate R_n and its true value λ; for

---

* In Hewitt's (1960) paper Canadian data yielded five pairs of (c, m) estimates ranging from $(30, 2.6)$ to $(42, 4.6)$.

example, $L(R_n - \lambda)$ could be $|R_n - \lambda|$. Optimality is achieved by minimization, with respect to R_n, of

$$\sum_{k=1}^{n} \int dU(\lambda) \int^{(k)} \int L(R_k - \lambda) \, dP(x_1 \mid \lambda) \, dP(x_2 \mid \lambda) \cdots dP(x_k \mid \lambda),$$

where all integrals are to be extended over the whole range of the corresponding variables; but, since

$$P(\lambda \mid x_1, x_2, \ldots, x_k) = \frac{U(\lambda) \prod_{j=1}^{k} dP(x_j \mid \lambda)}{\int \prod_{j=1}^{k} dP(x_j \mid \lambda) \, dU(\lambda)},$$

the expression to be minimized is equivalent to

$$\sum_{k=1}^{n} \int^{(k)} \int L(R_k - \lambda) \, dP(\lambda \mid x_1, x_2, \ldots, x_k) \int \prod_{j=1}^{k} dP(x_j \mid \lambda) \, dU(\lambda).$$

Since $L(R_k - \lambda) \geq 0$, this is achieved when

$$\int L(R_k - \lambda) \, dP(\lambda \mid x_1, x_2, \ldots, x_k)$$

is a minimum; for example, if we choose $L(R_k - \lambda) = (R_k - \lambda)^2$, the optimal "construction rule" $R_k{}^0$ is determined by minimizing

$$\int (R_k - \lambda)^2 \, dP(\lambda \mid x_1, x_2, \ldots, x_k),$$

namely, from

$$R_k{}^0 = \int \lambda \, dP(\lambda \mid x_1, x_2, \ldots, x_k),$$

the posterior mean of λ.

The experience-rating procedures described above require that an assumption be made about the form of $U(\cdot)$ and an initial estimate of its parameter values. An attempt to develop a nonparametric rating procedure is Simberg's (1964) suggestion that the net premium for the $(t + 1)$th contract year $\kappa(t + 1)$ $(t = 0, 1, 2, 3, \ldots)$ should be based on the claims of the prior year X_t and the preceding premium $\kappa(t)$ by means of the linear relation

$$\kappa(t + 1) = \alpha X_t + (1 - \alpha) \kappa(t), \qquad \alpha < 1 \text{ independent of } t. \tag{3.45}$$

This implies that

$$\begin{aligned} \kappa(t + 1) &= \alpha X_t + (1 - \alpha)[\alpha X_{t-1} + (1 - \alpha) \kappa(t - 1)] \\ &= \cdots \\ &= \alpha \sum_{j=0}^{t-1} (1 - \alpha)^j X_{t-j} + (1 - \alpha)^t \kappa(1). \end{aligned}$$

Because the greater weight is given to more recent experience, there is an implied assumption that there has been a time trend in the d.f. of X_t.

Neglecting this and assuming that the d.f. of X has remained invariant throughout the contract, we get

$$\mathcal{E} \, \kappa(t + 1) = \alpha\kappa_1 \frac{1 - (1 - \alpha)^t}{1 - (1 - \alpha)} + (1 - \alpha)^t \, \kappa(1)$$

$$= \kappa_1 + (1 - \alpha)^t\{\kappa(1) - \kappa_1\} \to \kappa_1 \quad \text{as} \quad t \to \infty. \quad (3.46)$$

Thus the system is consistent no matter what initial premium $\kappa(1)$ is charged. It is unbiased only when $\kappa(1) = \kappa_1$. Further

$$\mathcal{V} \, \kappa(t + 1) = \sum_{j=0}^{t-1} \alpha^2(1 - \alpha)^{2j} \mathcal{V} X = \frac{\alpha}{2 - \alpha} \, \kappa_2[1 - (1 - \alpha)^{2t-2}] \quad (3.47)$$

and this is always less than κ_2. Hence the system has a graduating effect on the series of aggregate claims.

A more satisfactory approach is that of Franckx (1964b) who proposed to make α dependent on t by putting $\alpha = 1/(t + \beta)$ with $\beta > 0$. Then

$$\kappa(t + 1) = \kappa(t) - \frac{\kappa(t) - X_t}{t + \beta} \quad (3.48)$$

$$= \frac{1}{t + \beta}\left[\beta \, \kappa(1) + \sum_{j=1}^{t} X_j\right], \quad (3.49)$$

so that

$$\mathcal{E} \, \kappa(t + 1) = \frac{1}{t + \beta} [\beta \, \kappa(1) + t\kappa_1]$$

and

$$\mathcal{V} \, \kappa(t + 1) = \frac{t}{(t + \beta)^2} \, \kappa_2,$$

with similar remarks about consistency and graduation. Franckx points out that (3.48) shows the corrective nature of the procedure, whereas (3.49) shows that a "weight" of β "years" is given to the first premium and unit weights to every annual X-value; $\beta = 0$ thus corresponds to no initial information and $\beta \to \infty$ means that experience is not allowed to influence the premium. If β is chosen fairly high, the tendency to "self-insurance" is damped down.

RECLASSIFICATION OF INDIVIDUAL CONTRACTS INTO "GOOD" AND "BAD" RISKS

The simplest type of heterogeneity of risks is the assumption that $U(\cdot)$ is a step-function with two unequal steps. Specifically, we write ($\lambda_1 < \lambda_2$)

$$U(\lambda_1 - 0) = 0, \qquad U(\lambda_1) = p, \qquad U(\lambda_2 - 0) = p, \qquad U(\lambda_2) = 1.$$

The problem is to classify risks into one or another of these categories on the basis of their claim records. If we assume that the distribution of individual claim amounts Y is independent of risk classification, we are concerned only with the numbers of claims an individual suffers in successive contract years.

In Chapter 2 we saw an illustration (Tröbliger, 1961, cited in Chapter 2) from the automobile accident field in which this dichotomy of risks provided a good fit of accident distribution. The estimates of the three parameters were $\hat{\lambda}_1 = 0.1089$, $\hat{\lambda}_2 = 0.7$ and $\hat{p} = 0.9403$. Let us consider, then, the problem of deciding whether a given contractholder has a Poisson λ-value of 0.1 or 0.7 based solely on his growing accident record and the assumption of invariance of λ. In the usual manner we reject the hypothesis that $\lambda = 0.1t$, where t is the number of years' records available, if n, the number of accident claims, exceeds a value n_0 such that

$$\mathfrak{F}\{n \geq n_0 \mid \lambda = 0.1t\} = \sum_{n=n_0}^{\infty} e^{-0.1t}\frac{(0.1t)^n}{n!} = \alpha, \qquad (3.50)$$

where α is small, say $\alpha < 0.05$. Now, if we reject $\lambda = 0.1t$, we must, on the dichotomy assumption, accept $\lambda = 0.7t$ so that the probability of correct acceptance is

$$\mathfrak{F}\{n \geq n_0 \mid \lambda = 0.7t\} = \sum_{n=n_0}^{\infty} e^{-0.7t}\frac{(0.7t)^n}{n!}. \qquad (3.51)$$

Table 3.3, extracted from the General Electric (1962) Tables of Poisson, shows the values of these two probabilities for small values of t and suitably

Table 3.3

t	Reject $\lambda = 0.1t$ If $n \geq n_0$, Where n_0 Equals	(3.50)	(3.51)
1	1	0.095	0.503
	2	0.005	0.156
2	1	0.181	0.753
	2	0.018	0.408
3	1	0.259	0.878
	2	0.037	0.620
4	2	0.062	0.769
	3	0.008	0.530
5	2	0.090	0.864
	3	0.014	0.679
6	2	0.122	0.922
	3	0.023	0.790
7	2	0.156	0.956
	3	0.034	0.867
8	2	0.191	0.976
	3	0.047	0.918

chosen n. Except for contract periods as long as seven or eight years, the results are rather discouraging—and, of course, would be even more ambiguous if the ratio λ_2/λ_1 were less than 7 (Gürtler, 1965). Thus, for example, after eight years any policyholder who is assigned to the "bad" category with $\lambda_2 = 0.7$ because he has suffered three or more accidents has a 92% chance of being correctly rated and there is only a 5% chance that he is a "good" risk with $\lambda_1 = 0.1$. On the other hand the use of a "one-claim" cirterion after three years to classify a contract as "good" means that 26% of "good" risks are misclassified as "bad," whereas 12% of "bad" risks are misclassified as "good." Since this is likely to drive the "good" risks who have been misclassified to a more accommodating insurance company, we may prefer to rate as "bad" only those contracts that have made two or more claims in three years: the result would then be that only 4% of "good" risks are classified as "bad" but 38% of "bad" risks are considered "good." It is obvious that such an approach to experience rating would be an administrative nightmare.

Instead of classifying a risk as "good" or "bad" and charging the appropriate premium after one or two years' experience, we may charge a graduated scale of premiums, depending on the number of claims made during the previous t years (Tröbliger, 1961; cited in Chapter 2). On the basis of the assumption of a dichotomy of risks, relation (3.38) becomes

$$\mathcal{E}(L, 1 \mid n, t) = \frac{1}{t}(n + 1)\frac{p_{n+1}(t)}{p_n(t)}$$

$$= \frac{1}{t}\frac{pe^{-\lambda_1 t}(\lambda_1 t)^{n+1} + (1 - p)e^{-\lambda_2 t}(\lambda_2 t)^{n+1}}{pe^{-\lambda_1 t}(\lambda_1 t)^n + (1 - p)e^{-\lambda_2 t}(\lambda_2 t)^n}$$

and this would be the premium to charge in the $(t + 1)$th year per unit of expected claim value. Furthermore, the expected proportion of contractholders who pay the premium corresponding to n previous claims is

$$p_n(t) = pe^{-\lambda_1 t}\frac{(\lambda_1 t)^n}{n!} + (1 - p)e^{-\lambda_2 t}\frac{(\lambda_2 t)^n}{n!} \ ;$$

for example, with $\lambda_1 = 0.15$, $p = 0.9\dot{3}$, and $\lambda_2 = 0.9$, we find, from a table provided by Tröbliger, the following premium rates per unit claim value for the $(t + 1)$th year based on n claims during the preceding t years. In parentheses immediately below each premium in Table 3.4 is shown the proportion of contractholders of that "generation" who are expected to pay that premium. The tendency toward the underlying dichotomy is fairly slow; at the end of 10 years 91.9% of the individual contracts with four claims or less are paying premiums of 0.187 or below and 5.5% with seven claims or more are paying premiums of 0.838 and over. However, the

Table 3.4

Number of Years t	Number of Claims n									
	0	1	2	3	4	5	6	7	8	9
1	0.174 (0.830)	0.276 (0.145)	0.561 (0.020)	0.810 (0.004)	0.883 (0.001)	0.896	0.900	0.900		
2	0.162 (0.702)	0.215 (0.227)	0.423 (0.049)	0.731 (0.014)	0.865 (0.005)	0.894 (0.002)	0.899 (0.001)	0.900		
3	0.156 (0.600)	0.182 (0.280)	0.310 (0.077)	0.614 (0.024)	0.830 (0.011)	0.887 (0.005)	0.898 (0.002)	0.900 (0.001)	0.900	
4	0.153 (0.514)	0.166 (0.314)	0.245 (0.104)	0.476 (0.033)	0.766 (0.016)	0.874 (0.009)	0.896 (0.006)	0.899 (0.003)	0.900 (0.001)	
5	0.151 (0.442)	0.157 (0.334)	0.193 (0.131)	0.350 (0.042)	0.664 (0.018)	0.847 (0.012)	0.891 (0.009)	0.898 (0.006)	0.900 (0.003)	0.900 (0.002)
6	0.151 (0.380)	0.154 (0.343)	0.171 (0.158)	0.260 (0.054)	0.530 (0.021)	0.795 (0.013)	0.880 (0.011)	0.897 (0.008)	0.899 (0.005)	0.900 (0.003)
7	0.150 (0.327)	0.152 (0.344)	0.160 (0.182)	0.206 (0.068)	0.395 (0.025)	0.708 (0.014)	0.859 (0.011)	0.893 (0.010)	0.899 (0.008)	0.900 (0.005)
8	0.150 (0.281)	0.151 (0.338)	0.155 (0.204)	0.178 (0.084)	0.290 (0.030)	0.584 (0.014)	0.819 (0.011)	0.885 (0.010)	0.897 (0.009)	0.900 (0.007)
9	0.150 (0.242)	0.150 (0.327)	0.152 (0.221)	0.163 (0.101)	0.223 (0.037)	0.446 (0.015)	0.747 (0.010)	0.869 (0.010)	0.895 (0.009)	0.899 (0.008)
10	0.150 (0.208)	0.150 (0.312)	0.151 (0.235)	0.156 (0.118)	0.187 (0.046)	0.326 (0.017)	0.636 (0.009)	0.838 (0.009)	0.889 (0.009)	0.898 (0.009)

advantage of this system over the two-premium dichotomy is that individuals charged in the neighborhood of 0.9, namely six times the "good" rate of 0.15, may judge from their own record that it is justified. In fact, certain of them are "good" risks who have been unlucky!

It is interesting to compare the foregoing with the premiums that would emerge if rerating were to be based only on the number of claim-free years out of t. If we write $e^{-\lambda_j} = q_j$ $(j = 1, 2)$, the expected value of L, the number of claims in the $(t + 1)$th year, knowing that m out of the t previous years have been claim-free, is

$$\frac{\sum_{l=0}^{\infty} l \left[p \frac{\lambda_1^{\,l}}{l!} q_1^{m+1}(1 - q_1)^{t-m} + (1 - p) \frac{\lambda_2^{\,l}}{l!} q_2^{m+1}(1 - q_2)^{t-m} \right]}{pq_1^{\,m}(1 - q_1)^{t-m} + (1 - p)q_2^{\,m}(1 - q_2)^{t-m}}$$

$$= \frac{p\lambda_1 q_1^{\,m}(1 - q_1)^{t-m} + (1 - p)\lambda_2 q_2^{\,m}(1 - q_2)^{t-m}}{pq_1^{\,m}(1 - q_1)^{t-m} + (1 - p)q_2^{\,m}(1 - q_2)^{t-m}}.$$

The expected proportion of contractholders with m claim-free years is

$$p\binom{t}{m} q_1^{\,m}(1 - q_1)^{t-m} + (1 - p)\binom{t}{m} q_2^{\,m}(1 - q_2)^{t-m}.$$

Table 3.5, based on $\lambda_1 = 0.15$, $p = 0.93$, and $\lambda_2 = 0.9$, shows that the claim-free procedure is slower than the aggregate-claims method to achieve the desired dichotomy in premiums. After 10 years 90.0% of the contractholders are paying premiums of 0.171 or less, and 4.1% are paying 0.866 or more. This is not much worse than the aggregate-claims procedure, though it will be noticed that the "bad" risks do not pay the appropriate higher premiums so early in the life of their contracts.

We mention that Tröbliger's paper (loc. cit.) also considers the problem of the premium to be charged if a dividend of $100 p_m \%$ is returned to the contractholder after m claim-free years. A less detailed derivation had previously been given in Delaporte (1959) and in Sousselier (1960).

An example of the segregation of "bad" risks applied to the case in which $U(\cdot)$ is a gamma distribution is found in Arbous and Sichel (1954). Concerned with employee absences (instead of accidents) in two nonoverlapping equilong intervals, the authors write the bivariate distribution (2.30) in the form

$$p_{nl}(1) = \left(\frac{c}{c + 2} \right)^m \frac{\Gamma(m + n + l)}{\Gamma(m)n!\, l!} \left(\frac{1}{c + 2} \right)^{n+l}$$

and derive the negative binomial array distribution

$$p_{l|n}(1) = \left(1 - \frac{1}{c + 2} \right)^{n+m} \frac{\Gamma(m + n + l)}{\Gamma(m + n)l!} \left(\frac{c}{c + 2} \right)^l$$

Table 3.5

Number of Years t	Number of Claim-Free Years m										
	10	9	8	7	6	5	4	3	2	1	0
1										0.174 (0.830)	0.325 (0.170)
2									0.162 (0.702)	0.244 (0.256)	0.573 (0.042)
3								0.156 (0.600)	0.198 (0.309)	0.435 (0.075)	0.785 (0.016)
4							0.153 (0.514)	0.173 (0.342)	0.318 (0.104)	0.692 (0.031)	0.869 (0.009)
5						0.151 (0.442)	0.161 (0.362)	0.240 (0.131)	0.564 (0.042)	0.838 (0.018)	0.893 (0.005)
6					0.151 (0.380)	0.155 (0.371)	0.195 (0.159)	0.426 (0.051)	0.780 (0.024)	0.884 (0.012)	0.898 (0.003)
7				0.150 (0.327)	0.153 (0.371)	0.172 (0.185)	0.312 (0.062)	0.684 (0.027)	0.868 (0.018)	0.896 (0.008)	0.900 (0.002)
8			0.150 (0.281)	0.151 (0.365)	0.161 (0.209)	0.236 (0.075)	0.555 (0.029)	0.835 (0.020)	0.892 (0.014)	0.899 (0.006)	0.900 (0.001)
9		0.150 (0.242)	0.151 (0.353)	0.155 (0.230)	0.193 (0.091)	0.417 (0.032)	0.775 (0.020)	0.884 (0.017)	0.898 (0.010)	0.900 (0.004)	0.900 (0.001)
10	0.150 (0.208)	0.150 (0.337)	0.152 (0.246)	0.171 (0.109)	0.305 (0.038)	0.677 (0.020)	0.866 (0.017)	0.896 (0.014)	0.900 (0.008)	0.900 (0.002)	0.900 (0.000)

so that

$$\Im\{L \geq l_0 \mid n\} = 1 - \left(1 - \frac{1}{c+2}\right)^{n+m} \frac{1}{\Gamma(m+n)} \sum_{l=0}^{l_0-1} \frac{\Gamma(m+n+l)}{l!} \left(\frac{c}{c+2}\right)^l.$$

(3.52)

Having estimated m and c from 318 absences in 1947 as 1.7164 and 0.34787, respectively, l_0 was chosen as 6 and (3.52) used to calculate the probabilities of having six or more absences in 1948 (Table 3.6).

Table 3.6

Number of Absences in 1947 (n)	Probability of Six or More Absences in 1948
6	0.475
7	0.568
8	0.652
9	0.726
10	0.786
11	0.840
12	0.881
13	0.911
14	0.931
15	0.948
16	0.961
17	0.971
18	0.979
19	0.985
20	0.990

By use of these probabilities the expected number of persons with six or more absences in 1948 was 55; the actual number was 52, and there was good agreement over the whole range of n-values. We note that if an expectation of six absences in a year is considered enough to classify a person as "bad" the observation of 15 absences in a year would make it 95% certain that the individual was "bad." A similar procedure for identifying children with high accident liabilities was used by Mellinger et al. (1965, cited in Chapter 2).

The foregoing techniques may be used to answer the question when to refuse insurance altogether. Haight (1965) has shown that when $U(\cdot)$ is a gamma distribution, so that $p_n(t)$ is negative binomial, the effect of eliminating individuals with $n \geq N$ is that the number of accidents L in a subsequent period s has pgf

$$G_H(z) = \left[\frac{c}{c+s(1-z)}\right]^m \frac{B_{(s+c-sz)/(t+s+c-sz)}(m, N)}{B_{c/(t+c)}(m, N)},$$

(3.53)

using the standard incomplete beta notation. Although no numerical results are available, an interesting fact about this distribution is that the negative binomial mean of sm/c for an interval of length s has been reduced to

$$\frac{sm}{c} - \frac{c^{m-1}t^{N}s}{(t+c)^{m+N}B_{c/(c+t)}(m,N)}. \tag{3.54}$$

The problem of reducing the mean number of accidents by selecting only individuals with $n = 0$ (Bhattacharya, 1967) and that of increasing the probability of having no accidents at all—a matter of great importance, for example, in the selection of test pilots—by excluding individuals with a heavy record of minor accidents (Bates and Neyman, 1954, cited in Chapter 2) have been considered in the literature.

SEQUENTIAL RATING BY MARKOV CHAIN THEORY

A particular type of "contagion" is allowed for in the assumption that the probability of having l claims in a contract year depends only on the number k of claims under that contract during the preceding year and is not a function of the numbers of claims occurring in prior years. It is as if the contractholder's claims during a given year were made with an eye to the number he made during the preceding year because his memory is too short to extend to prior years. Since the theory is usually developed on the basis of "states" (rather than numbers of claims) numbered $1, 2, 3, \ldots$ (instead of $0, 1, 2, \ldots$), we write $j = l - 1$ and $i = k - 1$ in the following presentation:

If r is the largest value assumed by i or j $(i, j = 1, 2, 3, \ldots, r)$ and p_{ij}, independent of t, represents the probability of moving from state i at epoch t to state j at epoch $t + 1$, the $r \times r$ matrix of probabilities

$$\mathbf{P} = ((p_{ij})), \qquad \sum_{j=1}^{r} p_{ij} = 1, \tag{3.55}$$

defines a finite Markov chain. If r is infinite, the chain is "denumerable," but "finite" theory (Feller, 1968; Cox and Miller, 1965) is sufficient for practical applications when r seldom exceeds 10.

We refer the reader to the cited references for development of the formulas we summarize below. Write $\mathbf{p}^{(t)}$ for the *row* vector of absolute probabilities

$$\{p_1^{(t)}p_2^{(t)}p_3^{(t)} \cdots p_r^{(t)}\}, \qquad \sum_{j=1}^{r} p_j^{(t)} = 1;$$

namely, the probabilities of being in state $j(j = 1, 2, \ldots, r)$ after t $(t = 1, 2, 3, \ldots)$ time intervals have passed from an initial epoch $t = 0$

when the row vector was $\mathbf{p}^{(0)}$. Then, since

$$p_j^{(t)} = \sum_{i=1}^{r} p_i^{(t-1)} p_{ij}, \qquad t = 1, 2, 3, \ldots,$$

we have

$$\mathbf{p}^{(t)} = \mathbf{p}^{(t-1)}\mathbf{P} = (\mathbf{p}^{(t-2)}\mathbf{P})\mathbf{P} = \mathbf{p}^{(t-2)}\mathbf{P}^2 = \cdots$$

$$= \mathbf{p}^{(0)}\mathbf{P}^t, \tag{3.56}$$

which shows that $\mathbf{p}^{(t)}$ is actually a function of t.

If $p_{ij}^{(t)}$ is the probability of moving from state i to state j in t time intervals, clearly

$$p_{ij}^{(2)} = \sum_{k=1}^{r} p_{ik} p_{kj} = \text{the } (i, j) \text{ element in } \mathbf{P}^2,$$

$$p_{ij}^{(3)} = \sum_{k=1}^{r} p_{ik}^{(2)} p_{kj} = \text{the } (i, j) \text{ element in } \mathbf{P}^3,$$

and, generally,

$$p_{ij}^{(t)} = \text{the } (i, j) \text{ element in } \mathbf{P}^t, \tag{3.57}$$

which is again a function of t.

If the series of absolute probability distributions $\mathbf{p}^{(0)}$, $\mathbf{p}^{(1)}$, $\mathbf{p}^{(2)}, \ldots,$ converges to a limiting distribution $\mathbf{p}$ so that $\lim_{t \to \infty} \mathbf{p}^{(t)} = \mathbf{p}$, the Markov chain is called ergodic, and since

$$\mathbf{p}^{(t+1)} = \mathbf{p}^{(t)}\mathbf{P},$$

$$\mathbf{p} = \mathbf{p}\mathbf{P}, \quad \text{with} \quad \sum_{j=1}^{r} p_j = 1. \tag{3.58}$$

When r is small, the calculation of $\mathbf{P}^t$ can be carried out directly but can become tedious for larger r and for large t-values. Because of the general availability of computer programs that calculate the so-called eigenvalues and eigenvectors of a square nonsymmetric matrix $\mathbf{P}$, the following formulas are of great practical utility:

Suppose that the r roots of the characteristic equation

$$|\mathbf{P} - \lambda\mathbf{I}| = 0 \tag{3.59}$$

are distinct and write them as the diagonal matrix

$$\Lambda = \begin{bmatrix} \lambda_1 & 0 & \cdots & 0 \\ 0 & \lambda_2 & \cdots & 0 \\ \cdots & \cdots & \cdots & \cdots \\ 0 & 0 & \cdots & \lambda_r \end{bmatrix}. \tag{3.60}$$

Then there exists a nonsingular matrix $\mathbf{H}$ such that

$$\mathbf{P} = \mathbf{H}\Lambda\mathbf{H}^{-1}, \tag{3.61}$$

where

$$\mathbf{H} = \{\mathbf{x}_1\mathbf{x}_2 \cdots \mathbf{x}_r\} \quad \text{and} \quad \mathbf{H}^{-1} = \{\mathbf{y}_1\mathbf{y}_2 \cdots y_r\}$$

are called the left and right eigenvectors of $\mathbf{P}$, respectively. Now, on squaring (3.61),

$$\mathbf{P}^2 = (\mathbf{H}\Lambda\mathbf{H}^{-1})(\mathbf{H}\Lambda\mathbf{H}^{-1}) = \mathbf{H}\Lambda\Lambda\mathbf{H}^{-1} = \mathbf{H}\Lambda^2\mathbf{H}^{-1}$$

and, generally,

$$\mathbf{P}^t = \mathbf{H}\Lambda^t\mathbf{H}^{-1}, \qquad t = 1, 2, 3, \dots, \tag{3.62}$$

where

$$\Lambda^t = \begin{bmatrix} \lambda_1{}^t & 0 & \cdots & 0 \\ 0 & \lambda_2{}^t & \cdots & 0 \\ \multicolumn{4}{c}{\dotfill} \\ 0 & 0 & \cdots & \lambda_r{}^t \end{bmatrix}.$$

Hence (3.62) is easily calculated once Λ, $\mathbf{H}$, and $\mathbf{H}^{-1}$ are provided by a computer.

The idea of applying the finite Markov chain model to automobile accident claims was first adumbrated by Fréchet (1959) and the theory restated by Franckx (1961a). The latter author was able to analyze three years' accident data in a note (1961b) published the same year; 966 French contracts classified as "Tous risques modernes" out of 1513 contracts initially written were in force at the end of the three-year period. By using only the data for these 966 contracts the row vectors of the transition matrix $\hat{p}_{ij}$, based on the "transitions" between the first and second years, are as follows (in which i and j are one more than the number of claims made).

i	1	j 2	3	
1	0.7108	0.2081	0.0811	
2	0.5127	0.3164	0.1709	$= \hat{\mathbf{P}}$ say
3	0.4194	0.3871	0.1935	
$\hat{\mathbf{p}}^{(1)}$	0.58696	0.28468	0.12836	

The corresponding matrix for the second to third year is

	1	2	3
1	0.74497	0.19295	0.06208
2	0.6245	0.2648	0.1107
3	0.41026	0.30769	0.28205
$\hat{\mathbf{p}}^{(2)}$	0.6170	0.2619	0.1211
$\hat{\mathbf{p}}^{(3)}$	0.6729	0.2257	0.1014

A check of the applicability of the Markov-chain model would have been obtained by multiplying the two transition matrices above and comparing the result with the matrix of transition probabilities between the first and third year, namely $\left(\left(\hat{p}_{ij}^{(2)}\right)\right)$. Unfortunately the latter estimate is not available from the published figures. The best we can do is to use the first matrix, which we have called $\hat{\mathbf{P}}$, to produce a "forecast" vector $\hat{\mathbf{p}}^{(3)}$. Thus

$$\hat{\mathbf{p}}^{(1)}\hat{\mathbf{P}}^2 = \{0.58696 \quad 0.28468 \quad 0.12836\} \begin{bmatrix} 0.6459 & 0.2652 & 0.1089 \\ 0.5983 & 0.2730 & 0.1287 \\ 0.5777 & 0.2847 & 0.1376 \end{bmatrix}$$

$$= \{0.6236 \quad 0.2582 \quad 0.1182\},$$

compared with the actual vector estimate

$$\hat{\mathbf{p}}^{(3)} = \{0.6729 \quad 0.2257 \quad 0.1014\}.$$

There is certainly enough general similarity to say that the Markov-chain assumptions are not unreasonable.

Let us now consider the rate of convergence of $\mathbf{p}^{(l)}$ to a limit. Using the aggregate of the two years "transitions," we obtain a new matrix estimate:

$$\hat{\hat{\mathbf{p}}}$$

$$\begin{bmatrix} 0.7283 & 0.2003 & 0.0714 \\ 0.5663 & 0.2917 & 0.1420 \\ 0.41494 & 0.34855 & 0.23651 \end{bmatrix}.$$

It follows that

$$\hat{\hat{\mathbf{P}}}^2 \qquad\qquad \hat{\hat{\mathbf{P}}}^3$$

$$\begin{bmatrix} 0.6735 & 0.2292 & 0.0973 \\ 0.63655 & 0.24801 & 0.11544 \\ 0.5977 & 0.2672 & 0.1351 \end{bmatrix} \qquad \begin{bmatrix} 0.6607 & 0.2357 & 0.1036 \\ 0.65196 & 0.24008 & 0.10796 \\ 0.6427 & 0.2447 & 0.1126 \end{bmatrix}.$$

$$\hat{\hat{\mathbf{P}}}^4 \qquad\qquad \hat{\hat{\mathbf{P}}}^5$$

$$\begin{bmatrix} 0.6577 & 0.2372 & 0.1051 \\ 0.6556 & 0.2383 & 0.1061 \\ 0.6534 & 0.2394 & 0.1072 \end{bmatrix} \qquad \begin{bmatrix} 0.6570 & 0.2376 & 0.1054 \\ 0.6565 & 0.2378 & 0.1057 \\ 0.6560 & 0.2381 & 0.1059 \end{bmatrix}.$$

Using only the initial vector estimate $\mathbf{p}^{(1)}$, we obtain

$$\hat{\mathbf{p}}^{(1)}\hat{\hat{\mathbf{P}}} = \hat{\hat{\mathbf{p}}}^{(2)}, \qquad\qquad \hat{\mathbf{p}}^{(1)}\hat{\hat{\mathbf{P}}}^2 = \hat{\hat{\mathbf{p}}}^{(3)}, \qquad\qquad \hat{\mathbf{p}}^{(1)}\hat{\hat{\mathbf{P}}}^3 = \hat{\hat{\mathbf{p}}}^{(4)}$$

$$\{0.6420 \quad 0.2453 \quad 0.1127\} \quad \{0.6533 \quad 0.2394 \quad 0.1073\} \quad \{0.6559 \quad 0.2381 \quad 0.1060\},$$

$$\hat{\mathbf{p}}^{(1)}\hat{\hat{\mathbf{P}}}^4 = \hat{\hat{\mathbf{p}}}^{(5)}, \qquad\qquad \hat{\mathbf{p}}^{(1)}\hat{\hat{\mathbf{P}}}^5 = \hat{\hat{\mathbf{p}}}^{(6)},$$

$$\{0.6566 \quad 0.2378 \quad 0.1056\} \quad \{0.65o7 \quad 0.2377 \quad 0.1056\}.$$

The tendency toward the limiting distribution $\mathbf{p}$ is very rapid and essentially achieved by $\hat{\hat{\mathbf{p}}}^{(6)}$. What is more surprising is that $\hat{\hat{\mathbf{P}}}^5$ indicates that $p_{ij}^{(t)}$ is independent of i for quite small t-values. This means that history is soon of little value in assessing an individual's probability of having j accident claims in a year. This conforms with U.S. practice in which three-year histories are regarded as sufficient for rating automobile risks. It suggests that (3.45) with relatively large α would be preferable to (3.48) if a Markov chain were a reasonable hypothesis.

A much larger experience derived from automobile accidents suffered by 20,326 policyholders of the Nationwide General Insurance Company of

Table 3.7

Old Class	New Class (when different) as a Result of Having			
	One Claim	Two Claims	Three Claims	Four of More Claims
		During the Three Prior Contract Years[a]		
11				
10		11	11	11
9		10	11	11
8		9	9	11
7		9	9	11
6		7	9	11
5		7	9	11
4	5	7	9	11
3	5	7	9	11
2	5	7	9	11
1	5	7	9	11

[a] Or during the whole contract when this has been in force only one or two years.

Columbus, Ohio, has been analyzed by Molnar and Rockwell (1966). In this case p_{ij} was the probability of moving from dividend class i to dividend class j and was estimated (apparently) from one year's experience of the company. The classification of policyholders for dividend purposes was based on a "merit-rating" procedure and was as follows: at the commencement of the contract the contractholder is assigned to a dividend class, for example, 8, depending on criteria developed by the insurance company, and he moves downward by one class for every no-claim year he experiences. His movement in case of one or more claims in a year is determined by Table 3.7.

The authors showed that $\hat{\mathbf{p}}^{(30)}$ calculated from (3.56) was approximately equal to $\hat{\mathbf{p}}^{(\propto)} = \hat{\mathbf{p}}$. They applied the relation

$$\hat{\mathbf{p}}^{(0)}\hat{\mathbf{P}} = \hat{\mathbf{p}}^{(1)}.$$

to forecast the vector of absolute probabilities for the following year. The agreement with actual observations was considered to be good and thus justified the merit-rating scheme.

It may be mentioned that Grenander (1957) has examined the consequences of a "merit-rating" system in which all individuals have the same probability p of making a claim but a probability p_{ij} of moving from bonus class i to bonus class j in which p_{ij} is the (i, j) element of the matrix

$$((a_{ij}p + b_{ij})) \quad \text{with} \quad \sum_j a_{ij} = 0 \quad \text{and} \quad \sum_j b_{ij} = 1.$$

He considered the expected premium income and the expected outgo in an equilibrium situation on the assumption that general expenses are increased by a fixed amount to settle any claim. In particular, he showed that these two expected values may not be equal for all p but the near equality may be attained for a range of p-values. This paper and two by Avondo-Bodino (1965) and Ottaviani (1966), respectively, provide theoretical approaches to the problem of "dividend hunger," namely the case in which the policy-holder will pay his own accident costs if they are less than his prospective dividend if he makes no claim in the year. Both authors assume that all policyholders have the same invariant probability of suffering an accident.

EXPERIENCE RATING BY CREDIBILITY INDICES

Since 1918 the premiums in successive years under workmen's compensation insurance contracts in the United States have been recalculated or "experience rated" by a formula of the form

$$\hat{k}_1 = Zk_1 + (1 - Z)K_1, \tag{3.63}$$

where K_1 is the "manual" premium rate based on supposedly extensive data (and thus regarded as a "true" mean value), k_1 is the premium rate deduced from the observations of a single contract (or a small class of contracts) with its own (Poisson) hazard λ, and Z is the "weight" or "credibility index" of that contract. The derivation of this formula is found in Whitney's (1918) pathbreaking paper in which it is assumed, in effect, that claims are uniformly equal to p_1 (in the notation of Chapter 2). We may thus rewrite (3.63) as

$$\hat{\lambda} = Z\left(\frac{k_1}{p_1 t}\right) + (1 - Z)\left(\frac{K_1}{p_1 t}\right). \tag{3.64}$$

If n claims have been made under the single contract in an exposure time of t years (e.g., t individuals for one year), the posterior density of Λ, the hazard variable, is by (3.34) proportional to

$$e^{-\lambda t}(\lambda t)^n \, dU(\lambda),$$

where $U(\cdot)$ is the prior distribution function of individual contract hazards. Let us assume that Λ is distributed as $\mathcal{N}(a, \sigma^2)$; then the posterior density of Λ is proportional to

$$(\lambda t)^n \exp\left[-\lambda t - \frac{1}{2\sigma^2}(\lambda - a)^2\right] \propto \lambda^n \exp\left[\frac{-(\lambda - a + \sigma^2 t)^2}{2\sigma^2}\right]. \quad (3.65)$$

We follow Whitney (*loc. cit*) by choosing the mode of this distribution as the point-estimate of $\hat{\lambda}$.

Differentiating the logarithm of (3.65) with respect to λ and equating to zero, we obtain

$$\frac{n}{\hat{\lambda}} - t - \frac{1}{\sigma^2}(\hat{\lambda} - a) = 0$$

or

$$\hat{\lambda}^2 + (t\sigma^2 - a)\hat{\lambda} - n\sigma^2 = 0. \quad (3.66)$$

Writing

$$\hat{\lambda} = \frac{n\sigma^2 + a^2}{t\sigma^2 + a} \quad (3.67)$$

and substituting into (3.66), we find that the expected value of its left-hand-side becomes

$$\mathcal{E}\left[\frac{(n/t - a)^2}{(1 + a/\sigma^2 t)^2}\right] = \frac{\sigma^2 + a/t}{(1 + a/\sigma^2 t)^2} = \frac{\sigma^2}{1 + a/\sigma^2 t},$$

since

$$\mathcal{E}n^2 = \underset{\lambda n}{\mathcal{E}}\mathcal{E}(n^2/\lambda) = \underset{\lambda}{\mathcal{E}}(\lambda t + \lambda^2 t^2) = at + (\sigma^2 + a^2)t^2.$$

If we are prepared to assume that the spread of the distribution of Λ is relatively narrow but that $\sigma^2 t \gg a$, the above expression is small. Hence (3.67) is an approximate solution of (3.66) and we may write

$$\hat{\lambda} = \frac{t\sigma^2}{t\sigma^2 + a} \cdot \frac{n}{t} + \frac{a}{t\sigma^2 + a} \cdot a = Z\frac{n}{t} + (1 - Z)a, \quad (3.68)$$

where

$$Z = \frac{t\sigma^2}{t\sigma^2 + a} = \frac{\sigma^2}{\sigma^2 + a/t}. \quad (3.69)$$

This estimate has the form of (3.64), and we note that Z involves the variances of (a) the mean number of claims based on sampling t observations from a Poisson distribution with parameter a, and (b) the Normal distribution of the hazard variable Λ.

We have seen that the mean of the posterior distribution of Λ, when its prior distribution is gamma, namely the mean of (3.36), is (in the present notation where $a = m/c$ and $\sigma^2 = m/c^2$)

$$\hat{\lambda} = \frac{n\sigma^2 + a^2}{t\sigma^2 + a},$$

which, coincidentally, is the same as Whitney's estimate (3.67). Hence in this case also Z is given by (3.69) (A. L. Bailey, 1950).

We mention that in Whitney (1918) the number of claims n was supposed to be distributed binomially rather than according to Poisson, though the (approximate) results were similar. In Bailey (1950) the binomial was considered as an alternative to Poisson and the prior for Λ was supposed to be a beta distribution. The same assumptions were also made by Finetti (1964). This binomial hypothesis would appear to be less in accordance with the occurrence of claims in the casualty insurance business than a Poisson assumption.

Since the "manual" rate K_1 after division by p_1 was regarded as a reasonably accurate estimate of a, the immediate problem was to estimate σ^2. In the Whitney (1918) paper it is mentioned that trials with workmen's compensation data suggested that $\sigma^2 \propto a^{5/4}$, but the simpler alternative

$$Z = \frac{ta}{ta + k},\qquad(3.70)$$

where k is a constant "determined by judgment," was found to provide a good approximation. Thus Michelbacher (1918) found that the k for death and permanent total disability claims in Illinois was approximately 18×10^3, whereas for other types of claim it was 8×10^3. The reason for subdividing the claims in this manner was that a for the former type ranged from 10^{-4} to 10^{-2} and for the latter from 10^{-3} to $\frac{1}{2}$.

An important feature of Z determined from (3.69) or (3.70) is that as t, the exposure, increases Z approaches unity. Since the credibility indices Z were subject to state governmental review and to public criticism, it became essential to device a simple, "understandable" formula that would broadly approximate the truth. A relation for Z which has been widely utilized is

$$Z = \left[\frac{\min{(t, T)}}{T}\right]^{1/2},\qquad(3.71)$$

where T is the value of t which is supposed to give "full credibility," namely, $Z = 1$. Several papers (e.g., Perryman, 1932) have considered the problem of determining T on probabilistic grounds as the number of observations that make it almost certain that n/T is very close to a.

The subjective element in hazard estimation that was explicitly recognized from the outset had thus moved from ascribing a value to σ^2 in (3.69) to giving a value to k in (3.70) and finally to providing a value for T in (3.71).

An attempt to provide a theoretical justification for this square-root formula was later made by A. L. Bailey (1943).

Further examples of the formulation of relations for Z are found in Kormes (1952) and Hurley (1954). Hurley proposed a hyperbola for small t-values and a straight line for t-values in excess of a so-called "focal" value. In the discussion of this paper M. H. McConnell provided several alternatives including (part of) an ellipse and a parabola. Although recognizing that a subjective element may legitimately be introduced into experience rating, the writer wonders whether elaborate investigations into suitable mathematical forms for Z have not tended to obscure the fact that σ^2 of (3.69) could be estimated from actual data.

As an illustration of experience rating, using credibility indices, we may consider the calculation of Z for a given class of workmen's compensation risks in Massachusetts (Scammon, 1947). Since 1923 it had been recognized that an approximation to the variability of individual claim amounts would be achieved by a dichotomy consisting of small and large amounts, respectively. In Massachusetts this was effected pragmatically by calling all individual claims of less than \$400 "primary" claims with "substantial credibility" and by writing a claim $X > 400$ as

$$X = 400r + R, \qquad r = 0, 1, 2, 3, \ldots, \qquad R < 400;$$

X was then split into a "primary" portion, namely,

$$X_p = 400 \sum_{j=0}^{r-1} (\tfrac{2}{3})^j + R(\tfrac{2}{3})^r,$$

and an "excess" portion with "little credibility," namely,

$$X_e = X - X_p.$$

The National Council Actuarial Committee decided "after much consideration" that "full credibility" would be attained by 250 "primary" claims or by 500 "excess" claims and adopted the formula

$$Z = \left[\frac{\min(t, T)}{T} \right]^{2/3}. \qquad (3.72)$$

Thus, for example, in gear manufacturing and grinding in 1943–1944 there were 202 employers with a payroll of \$27,769,272, "primary" claims aggregating \$70,772 (0.0025 of the payroll), and "excess" claims by 27 employers aggregating \$19,442 (0.0007 of the payroll). The "primary" credibility index is thus $Z_p = (202/250)^{2/3} = 0.87$ and the "excess" $Z_e = (27/500)^{2/3} = 0.14$. The corresponding experience-rated "primary" premium for the class was accordingly

$$\hat{\kappa}_{1p} = 0.87 \times 0.0025 + 0.13 \times 0.0044 = 0.0027,$$

since the "manual" rate was $K_1 = 0.0044$. After a similar calculation for $\hat{\kappa}_{1e}$ the sum of $\hat{\kappa}_{1p}$ and $\hat{\kappa}_{1e}$ was the experience-rated premium for the class.

A substantial disadvantage of the foregoing procedures is that estimation of $P(\cdot)$, the d.f. of individual claim amounts, or of $F(\cdot)$, the d.f. of aggregate claims, has been replaced by the rather crude splitting of claims into "primary" and "excess" or, at best, into several size groups (a "multisplit" plan). This device was extended by R. A. Bailey (1961) to a "split" at every consecutive point on the dense set $0 < x < \infty$. The writer found this author's heuristic arguments difficult to understand, particularly in their reference to a "true" or estimated $F(\cdot)$. However, in this case also we are dubious about a derivation based on Z rather than on $P(\cdot)$, a, and σ^2.

REFERENCES

Ammeter, H. (1957). "Die Ermittlung der Risikogewinne im Versicherungswesen auf risikotheoretischer Grundlage." *Mitt. Verein. Schweiz. Versich. Mathr.*, **57**, 145–200. [This paper partly reproduced in (1957) "A rational experience rating technique for group insurance on the risk premium basis." *Trans. XV Intern. Cong. Actu.*, New York, **1**, 507–521.]

——— (1961). "Risikotheoretische Grundlagen der Erfahrungstarifierung." *Mitt. Verein. Schweiz. Versich. Mathr.*, **61**, 183–217. [A free translation of this 1961 article appears in (1960) "Stop loss cover and experience rating." *C.R. XVI Cong. Intern. Actu.*, Bruxelles, **1**, 649–661. (1963) "Experience rating: A new application of the collective theory of risk." *Astin Bull.*, **2**, 261–270.]

Arbous, A. G., and H. S. Sichel (1954). "New techniques for the analysis of absenteeism data." *Biometrika*, **41**, 77–90.

Avondo-Bodino, G. (1955). "Un'osservazione a proposito della introduzione del ristorno nell assicurazione R.C.A." *Gior. Ist. Ital. Attu.*, **28**, 147–153.

Bailey, A. L. (1942/1943). "Sampling theory in casualty insurance." *Proc. Casualty Actu. Soc.*, **29**, 50–95; **30**, 31–65.

——— (1950). "Credibility procedures: Laplace's generalization of Bayes' rule and the combination of collateral knowledge with observed data." *Proc. Casualty Actu. Soc.*, **37**, 7–23.

Bailey, R. A. (1961). "Experience rating reassessed." *Proc. Casualty Actu. Soc.*, **48**, 60–82.

Bailey, R. A., and L. R. Simon (1960). "Two studies in automobile insurance ratemaking." *Astin Bull.*, **1**, 192–217.

Benktander, G. (1953). "On the variation of the risk premium with the dimensions of the house within fire insurance." *Skand. Aktu. Tidskr.*, **36**, 203–214.

Bhattacharya, S. K. (1967). "A result on accident proneness." *Biometrika*, **54**, 324–325.

Bichsel, F. (1961). "Une méthode pour calculer une ristourne adéquate pour années sans sinistres." *Astin Bull.*, **1**, 106–112.

——— (1964). "Erfahrungs-Tarifierung in der Motorfahrzeughaftpflicht-Versicherung." *Mitt. Verein. Schweiz. Versich. Mathr.*, **64**, 119–130.

Bühlmann, H. (1964). "Optimale Pramienstufensysteme." *Mitt. Verein. Schweiz. Versich. Mathr.*, **64**, 193–214.

Conolly, B. W. (1955). "Unbiased premiums for stop-loss reinsurance." *Skand. Aktu. Tidskr.*, **38**, 127–134.

Cox, D. R., and H. D. Miller (1965). *The Theory of Stochastic Processes*. Methuen, London.

Delaporte, P. (1959). "Quelques problèmes de statistiques mathématiques posés par l'assurance automobile et le bonus pour non sinistre." *Bull. Trim. Inst. Actu. Franç.*, **58**, 87–102.

——— (1962). "Sur l'efficacité des critères de tarification de l'assurance contre les accidents d'automobiles." *Astin Bull.*, **2**, 84–95.

Delaporte, P. J. (1965). "Tarification du risque individuel d'accidents d'automobiles par la prime modelée sur le risque." *Astin Bull.*, **3**, 251–271.

Draper, N. R., and H. Smith (1966). *Applied Regression Analysis*. Wiley, New York.

Dropkin, L. B. (1960). "Automobile merit rating and inverse probabilities." *Proc. Casualty Actu. Soc.*, **47**, 37–40.

Dubois, P. (1937). "Contribution à l'étude le la réassurance." *C.R. XI Cong. Intern. Actu.*, Paris, **1**, 397–412.

Esscher, F. (1965). "Some problems connected with the calculation of stop loss premiums for large portfolios." *Skand. Aktu. Tidskr.*, **48**, 65–80.

Feller, W. (1968). *An Introduction to Probability Theory and Its Applications*. Vol. I. Wiley, New York.

Finetti, B. de (1964). "Sulla teoria della credibilità." *Gior. Ist. Ital. Attu.*, **27**, 219–231.

Fisher, R. A., and W. A. Mackenzie (1923). "Studies in crop variation. II. The manurial response of different potato varieties." *J. Agric. Sci.*, **13**, 311–320.

Franckx, E. (1961a). "Théorie du bonus: Conséquences de l'étude de Mr. le Professeur Fréchet." *Astin Bull.* **1**, 113–122.

——— (1961b). "Les assurances 'accident-automobile'." *C.R. XVI Cong. Intern. Actu.*, Brussels, **3**, 209–213.

——— (1964a). "On some essential properties of a tariff class." *Proc. Casualty Actu. Soc.*, **51**, 131–139.

——— (1964b). "The adaptation of motor insurance rates using secondary information." *Quarterly Letter*, *Jubilee* No. **2**, 39–45, Algemeene Reins. Co. [French translation in *Bull. Assoc. Roy. Actu. Belges*, **62**, 9–17.]

Fréchet, M. (1959) "Essai d'une étude des successions de sinistres considérés comme processus stochastique." *Bull. Trim. Inst. Actu. Franç.*, **58**, 67–85.

Fürst, D. (1965). "Formulation bayesienne du problème des valeurs extrêmes en relation à la réassurance en 'excédent de sinistres'." *Astin Bull.*, **3**, 153–162.

General Electric Company (1962). *Tables of the Individual and Cumulative Terms of Poisson Distribution*. Van Nostrand, Princeton, N.J.

Grenander, U. (1957). "Some remarks on bonus systems in automobile insurance." *Skand. Aktu. Tidskr.*, **40**, 180–197.

Gumbel, E. J. (1958). *Statistics of Extremes*. Columbia University Press, New York.

Gürtler, M. (1965). "Bonus or malus." *Astin Bull.*, **3**, 43–61.

Haight, F. A. (1965). "On the effect of removing persons with N or more accidents from an accident prone population." *Biometrica*, **52**, 298–300.

Hewitt, C. C., Jr. (1960). "The negative binomial applied to the Canadian merit rating plan for individual automobile risks." *Proc. Casualty Actu. Soc.*, **47**, 55–65.

Hooge, L. D'. (1965). "Théorie des valeurs extrêmes et la tarification de l' *excess of loss*." *Astin Bull.*, **3**, 163–177.

Hurley, R. L. (1954). "A credibility framework for gauging fire classification experience." *Proc. Casualty Actu. Soc.*, **41**, 161–175.

Jackson, P. H. (1953). "Experience rating." *Trans. Soc. Actu.*, **5**, 239–255.

Johnson, N. L. (1957). "Uniqueness of a result in the theory of accident proneness." *Biometrika*, **44**, 530–531.

———— (1967). "Note on a uniqueness relation in certain accident proneness models." *J. Amer. Statist. Assoc.*, **62**, 288–289.

Jung, J. (1965). "On the use of extreme values to estimate the premium for an excess of loss reinsurance." *Astin Bull.*, **3**, 178–184.

Keffer, R. (1929). "An experience rating formula." *Trans. Actu. Soc. Amer.*, **30**, 130–139, 593–611.

Kormes, M. (1952). "Note on experience rating credibility." *Proc. Casualty Actu. Soc.*, **39**, 98–99.

Lundberg, O. (1941). "On the importance of regarding risk premiums in voluntary sickness and accident insurance: a theoretical basis for regrading." *Ber. XII Intern. Kong. Versich. -Mathr.*, Lucerne, **2**, 543–554.

———— (1966). "Une note sur des systèmes de tarification basés sur des modèles du type Poisson composé." *Astin Bull.* **4**, 49–58.

Mehring, J. (1965). "Ein mathematisches Hilfsmittel für Statistik- und Tariffragen in der Kraftfahrtversicherung" (Anhang by C. Boehm). *Bl. Deuts. Gesell. Versich. Math.*, **7**, 111–140.

Michelbacher, G. F. (1918). "The practice of experience rating." *Proc. Casualty Actu. Soc.*, **4**, 293–324.

Molnar, D. E., and T. H. Rockwell (1966). "Analysis of policy movement in a merit-rating program: an application of Markov processes." *J. Risk. Ins.*, **33**, 265–276.

Ottaviani, R. (1966). "Sull' introduzione del ristorno nelle assicurazioni di risarcimento." *Gior. Ist. Ital. Attu.*, **29**, 97–113.

Perryman, F. S. (1932). "Some notes on credibility." *Proc. Casualty Actu. Soc.*, **19**, 65–84.

Pesonen, E. (1963). "A numerical method of finding a suitable bonus scale." *Astin Bull.*, **2**, 102–108.

Philipson, C. (1959). "A method of estimating grouped frequencies." *Skand. Aktu. Tidskr.*, **42**, 159–193.

Scammon, L. W. (1947), "Massachusetts workmen's compensation rate making: Primary-excess basis." *Proc. Casualty Actu. Soc.*, **34**, 17–49.

Seal, H. L. (1964). *Multivariate Statistical Analysis for Biologists.* Methuen, London.

———— (1968). "The use of multiple regression in risk classification based on proportionate losses." *Ber. XVIII Intern. Kong. Versich. Mathr.*, Munich, **2**, 659–664.

Simberg, H. (1964). "Individuelle Prämienregelung, eine Art des 'experience rating'." *Trans. XVII Intern. Cong. Actu.*, London, **3**, 650–657.

Solomon, L. (1948). "The analysis of heterogeneous mortality data." *J. Inst. Actu.*, **74**, 94–112.

Sousselier, J. (1960). "Bonus, franchise et surprimes pour sinistres." *Bull. Inst. Sci. Finan. Assur.*, No. **20**, 5–21.

Vajda, S. (1930). "Gewinnbeteiligung und Risikotheorie." *C.R. IX Cong. Intern. Actu.*, Stockholm, **2**, 254–275.

———— (1943). "The algebraic analysis of contingency tables." *J. Roy. Statist. Soc.*, **106**, 333–342.

———— (1946). "The analysis of variance of mortality rates." *J. Inst. Actu.*, **72**, 240–245.

———— (1947). "Statistical investigation of casualties suffered by certain types of vessels." *J. Roy. Statist. Soc. Suppl.*, **9**, 141–175.

———— (1951, 1955). "Analytical studies in stop-loss reinsurance." *Skand. Aktu. Tidskr.*, **34**, 158–175; **38**, 180–191.

Whitney, A. W. (1918). "The theory of experience rating." *Proc. Casualty Actu. Soc.*, **4**, 274–292.

CHAPTER 4

The Probability of Ruin of a Risk Business

We have already stated that when the risk reserve (capital and surplus) of an insurance company dwindles to zero or less the company is said to be ruined. While it continues in business the risk reserve is supposed to be increased by the risk-loaded premiums received from contractholders, and depleted by all claims paid. It is clear that if the risk reserve at a given moment is many times the largest sum-at-risk that could become payable ruin cannot occur in the near future unless an exceptionally large risk is accepted in the meantime. On the other hand, if the risk reserve is only equal to κ_1 plus about once the standard deviation of X, the claim outgo of a year, there is a good chance of ruin occurring within one or two years. It is the intermediate situation, in which the risk reserve is equal to κ_1 plus a fairly large multiple of the standard deviation of X, that has given rise to numerous theoretical treatments since the end of the last century. In all of these treatments the principal objective has been to calculate the probability of the ruin of a given risk business within a specified period of time, which at first was as short as one year and later extended indefinitely.

The theory thus developed is summarized in this chapter. It is based on the assumption that the parameters of the claim outgo distribution are known, even if conceptually they may be subject to secular changes. To the statistician what is missing is any consideration of the possibility that estimates made of these parameters may be biased in favor of the contractholders so that the aggregate net premiums charged may be less than κ_1. If this were to happen the risk business might find itself with a consistently decreasing risk reserve, and its response should be the recalculation of the parameter estimates rather than an increase in η_1, the risk loading added to κ_1. In either case, of course, the effect on the contractholders would be the same, namely an increase in their premiums, but the reasons for the increase are quite different and perhaps call for a reformulation of the problem of ruin. In particular, we note that a run of "bad luck" may continue indefinitely,

90

and the company that believes its proper recourse is to increased net premiums, or to a complete reassessment of its rating procedures, is deluding itself about its ability to stay in business.

Consider a risk business that is receiving a continuous stream of premiums from contractholders who are "covered" against risks that occur at random. The number of these contractholders may be regarded as a function N_τ of the time τ elapsed since an arbitrary time origin. Although the accession of new contractholders and subsequent contract cancellations may be regarded as chance occurrences, we consider N_τ as a given function of τ subject to abrupt upward and downward changes that occur in a known, deterministic manner. In particular, the occurrence of a claim (a supposedly random happening) is not assumed to reduce the then-value of N_τ by a unit. If the risk business is life insurance subject to risk premiums, this means, in effect, that $N_\tau \gg 1$.

If the probability of a claim at time τ is equal to $N_\tau \lambda_\tau \, d\tau + o(d\tau)$, relation (2.16) provides the probability of n claims occurring in the interval $(0, t)$, namely,

$$p_n(t) = \exp\left(-\int_0^t N_\tau \lambda_\tau \, d\tau\right) \frac{\left(\int_0^t N_\tau \lambda_\tau \, d\tau\right)^n}{n!}, \qquad n = 0, 1, 2, \ldots . \quad (4.1)$$

As indicated in Chapter 2 following relation (2.16), it is convenient to make a transformation of the time scale such that (4.1) can be written

$$p_n(t) = e^{-t} \frac{t^n}{n!}, \qquad n = 0, 1, 2, \ldots, \quad (4.2)$$

and this means that "time t" is the interval during which t claims are "expected." Note that the probability of a claim at time τ which was equal to $N_\tau \lambda_\tau \, d\tau$ is now simply $d\tau$. Observe, too, that if a large risk business "expects" 100,000 claims in a year of real time its probability of ruin within that period (namely, within 100,000 "time" units) is hardly distinguishable from its probability of ruin in an indefinitely long period. This remark shows why the problem of "eventual" ruin for a large company is essentially the same as its probability of ruin within a fairly short period of years.

We have arranged the material of this chapter as follows. We are concerned with the calculation of the probability that a risk business's risk reserve becomes zero or negative as a result of excess claims during a specified period of time (or "time") commencing immediately. Because it is historically important and aptly illustrates the techniques to be used, we commence with the probability of the eventual ruin (i.e., $t \to \infty$) of an insurance company, all of whose policies are for the same sum-at-risk h. In order to complete the solution, we have to calculate this probability for the special case in which the initial risk reserve is zero. The same problems are then considered under the limitation that nonoverlapping contracts, each of which lasts a uniform period, are undertaken sequentially (random walk).

The general problem of ruin within a given "time" t (namely, during a period when expected claims number t) is then solved directly and by means of Laplace transforms. These solutions are not easily adapted to numerical calculations and asymptotic expressions are given for the moments of T, the random variable of the "time" (i.e., the expected number of claims) to ruin, which is asymptotically Normal in distribution. Similar theory is then provided for a company writing only life annuity contracts.

We next turn to the theory for the case in which the interval (whether measured in real time units or in expected-claim units) is increased indefinitely. This historically preceded the more general theory and provides practical results for the large risk business. The Laplace transform of the probability of eventual ruin is now much simpler and has even been inverted explicitly in some special cases. The integral equation for this probability is an example of the general renewal equation, and the opportunity is taken to derive its solution directly and to provide the appropriate asymptotic theory that allows some of the earlier actuarial derivations to be materially simplified.

THE PROBABILITY OF ULTIMATE RUIN WITH EQUAL SUMS AT RISK

Let $Y(t)$ equal the aggregate of (positive) claims in the period of time $(0, t)$. Then, if $\eta_1 \, d\tau$ is the risk loading added to the premium $p_1 \, d\tau$ received in the interval of (transformed) time $(\tau, \tau + d\tau)$, the aggregate (algebraic) gain to the risk business up to time t is $(p_1 + \eta_1)t - Y(t)$, and we consider the risk reserve at time t, namely,

$$R(t) = R(0) + (p_1 + \eta_1)t - Y(t). \tag{4.3}$$

The plot of $R(t)$ against t consists of a straight line sloping upward from the point $(0, R(0))$ with slope $p_1 + \eta_1$ and subject to vertical downward jumps of random length Y at random intervals of time. Ruin occurs when one of these jumps causes the curve $R(t)$ to cross the time axis, and we are going to consider the probability of such a crossing before the end of time t where $t \to \infty$. [We refer the reader to Cramér (1955) and Arfwedson (1957) for discussion of the existence and uniqueness of a probability distribution of the stochastic process $Y(t)$ and of the ruin probabilities considered now and hereafter.]

We commence with the special case in which all sums at risk (or individual claim amounts) are equal to h, so that $p_1 = h$, and henceforth we shall write

$p_1 + \eta_1 \equiv \pi_1$. (In Chapter 3 π_1 was the risk premium payable in a single year of real time. Here π_1 is a *rate* of premium per expected claim.) Let us find the probability that the first claim occurs at time τ. It is equal to the product of the probabilities of the independent events (a) that no claim occurs in the interval $(0, \tau - d\tau)$—namely $e^{-\tau}$—and (b) that a claim then occurs in $(\tau - d\tau, \tau)$, namely $d\tau$. The required probability is thus $e^{-\tau} d\tau$, and we may express this by saying that under the Poisson assumption the density of interclaim periods is negative exponential.

We now write $\psi(w)$ for the probability of ultimate ruin, given that w is the initial risk reserve; that is, $R(0) = w$. Let $1 - \psi(w) = \phi(w)$. If we then move along the t-axis until the first claim occurs, the risk reserve will have increased to $w + \pi_1 t$ and the claim will reduce it to $w + \pi_1 t - h$ so that the probability of subsequent ruin is $\psi(w + \pi_1 t - h)$. Arguing in terms of nonruin instead of ruin, we may thus write

$$\phi(w) = \int_0^\infty e^{-\tau} \phi(w + \pi_1 \tau - h)\, d\tau \qquad (4.4)$$

and

$$\phi(w) = 0, \qquad w < 0,$$

but we must be prepared for a discontinuity at $w = 0$ and, as we shall see, $\phi(0)$ is only zero when $\eta_1 = 0$.

Equation 4.4 with $\pi_1 = 1$ and $\phi(0) = 1 - h$ goes back to Erlang (1909), who developed it in connection with delays in answering telephone calls. Suppose that calls arrive randomly with unit intensity (i.e., we choose the unit of time so that one call—or "claim"—is expected during any time unit) at an exchange manned by a single operator. The operator requires h time units to make a connection for a caller, and if he is thus engaged when another call arrives the new call waits on line until the operator is free. Several calls may arrive and form a queue while a single call is being connected.

Write $\phi(w)$ for the probability that the connection delay is less than (or equal to) w. In the long run the proportion of calls that find the operator engaged in making connections is $h:1$ (showing that $h < 1$ if the problem is to be soluble) and thus $\phi(0) = 1 - h$. Suppose that a caller rings the exchange τ time units after the previous caller—the probability of this is $e^{-\tau} d\tau$—and has to wait less than w time units for his connection. This caller thus has a probability $\phi(w)$ of a delay less than w; but in order that this delay may be possible the previous caller must have suffered a delay of less than $\tau + w - h$ time units. Hence (4.4) holds with $\pi_1 = 1$.

Erlang (*loc. cit.*) solved (4.4) by reducing it to a differential equation which he integrated. Another method is to recognize that $\phi(jh) - \phi(\overline{j - 1}h)$ is the probability that there will be j callers waiting in the queue and to derive a recurrence relation for this probability (Fry, 1928). A third method is to take the Laplace transform of both sides of (4.4). Repeating the definition of the Laplace (L.T.) of a function $g(x), 0 < x < \infty$, such that $\lim_{x \to \infty} e^{-sx} g(x) = 0$,

$$\mathcal{L}\{g(x)\} \equiv \gamma(s) = \int_0^\infty e^{-sx} g(x) \, dx,$$

where the integral is assumed to exist for $\Re(s) > \sigma_0 > 0$. [We notice that if x were values assumed by a random variable X and $M(\theta)$ its moment-generating function then $\gamma(s) = M(-s)$.] It can be proved that a unique $g(\cdot)$ corresponds to a given L.T.—though this does not necessarily mean we can determine $g(\cdot)$ explicitly, even though we can always approximate it numerically. Now, $w \geq 0$,

$$\phi(w) = \int_0^\infty e^{-\tau} \phi(w + \pi_1 \tau - h) \, d\tau$$

$$= \pi_1^{-1} \int_w^\infty e^{-(z-w)/\pi_1} \phi(z - h) \, dz$$

$$= \frac{e^{w/\pi_1}}{\pi_1} \int_w^\infty e^{-z/\pi_1} \phi(z - h) \, dz.$$

On taking the L.T. of both sides

$$\int_0^\infty e^{-sw} \phi(w) \, dw \equiv \Phi(s) = \pi_1^{-1} \int_0^\infty e^{-w(s-\pi_1^{-1})} \, dw \int_w^\infty e^{-z/\pi_1} \phi(z - h) \, dz$$

$$= \pi_1^{-1} \int_0^\infty e^{-z/\pi_1} \phi(z - h) \, dz \int_0^z e^{-w(s-\pi_1^{-1})} \, dw$$

$$= \pi_1^{-1} \int_h^\infty \phi(z - h) e^{-z/\pi_1} \frac{e^{-z(s-\pi_1^{-1})} - 1}{-(s - \pi_1^{-1})} \, dz$$

$$\text{since} \quad \phi(w) = 0, \qquad u < 0$$

$$= (\pi_1 s - 1)^{-1} \left[\int_h^\infty e^{-z/\pi_1} \phi(z - h) \, dz \right.$$

$$\left. - \int_h^\infty e^{-sz} \phi(z - h) \, dz \right]$$

$$= (\pi_1 s - 1)^{-1} [e^{-h/\pi_1} \Phi(\pi_1^{-1}) - e^{-sh} \Phi(s)]$$

$$= (\pi_1 s - 1 + e^{-sh})^{-1} e^{-h/\pi_1} \Phi(\pi_1^{-1}). \tag{4.5}$$

The last two factors on the right are constants. It is thus our objective to recognize the reciprocal preceding these constants as the L.T. of some "standard" function—if such a function does indeed exist explicitly. We thus proceed:

$$\Phi(s) = e^{-h/\pi_1} \, \Phi(\pi_1^{-1})(\pi_1 s - 1)^{-1}[1 + (\pi_1 s - 1)^{-1}e^{-sh}]^{-1}$$

$$= e^{-h/\pi_1} \, \Phi(\pi_1^{-1})(\pi_1 s - 1)^{-1} \sum_{j=0}^{\infty}(-1)^j e^{-shj}(\pi_1 s - 1)^{-j}$$

$$= e^{-h/\pi_1} \, \Phi(\pi_1^{-1}) \sum_{j=0}^{\infty} \frac{(-1)^j}{j!} \, e^{-shj}(\pi_1 s - 1)^{-j-1} \, \Gamma(j + 1)$$

$$= e^{-h/\pi_1} \, \Phi(\pi_1^{-1}) \sum_{j=0}^{\infty} \frac{(-1)^j}{j!} \, e^{-shj}(\pi_1 s - 1)^{-j-1} \int_0^{\infty} x^j e^{-x} \, dx$$

$$= e^{-h/\pi_1} \, \Phi(\pi_1^{-1}) \sum_{j=0}^{\infty} \frac{(-1)^j}{j!} \, e^{-shj} \int_0^{\infty} v^j e^{-v(\pi_1 s - 1)} \, dv$$

$$= e^{-h/\pi_1} \, \Phi(\pi_1^{-1}) \sum_{j=0}^{\infty} \frac{(-1)^j}{j!} \, \pi_1^{-1} \int_{hj}^{\infty} \left(\frac{w - hj}{\pi_1}\right)^j e^{(w-hj)/\pi_1 - sw} \, dw.$$

The integral last written is, of course, the L.T. of the function

$$\left(\frac{w - hj}{\pi_1}\right)^j e^{(w-hj)/\pi_1}, \qquad w \geq hj.$$

Inverting, we thus have for $w \geq 0$

$$\phi(w) = \pi_1^{-1} e^{-h/\pi_1} \, \Phi(\pi_1^{-1}) \sum_{j=0}^{[w/h]} \frac{(-1)^j}{j!} \left(\frac{w - hj}{\pi_1}\right)^j \exp\left(\frac{w - hj}{\pi_1}\right), \quad (4.6)$$

where $[x]$ means "greatest integer in x." On putting $w = 0$, we obtain

$$\phi(0) = \pi_1^{-1} e^{-h/\pi_1} \, \Phi(\pi_1^{-1}),$$

hence

$$\Phi(\pi_1^{-1}) = \pi_1 e^{h/\pi_1} \, \phi(0),$$

so that (4.6) becomes

$$\phi(w) = \phi(0) \sum_{j=0}^{[w/h]} \frac{(-1)^j}{j!} \left(\frac{w - hj}{\pi_1}\right)^j \exp\left(\frac{w - hj}{\pi_1}\right). \quad (4.7)$$

It is thus necessary to evaluate $\phi(0)$ before we can calculate $\phi(w)$ for $w > 0$.

In order to obtain an idea of the rate of convergence toward unity of the function $\phi(w)$, we may take $h = 9$ and $\pi_1 = 10$. Anticipating the formula we shall derive for $\phi(0)$, we state that in this case $\phi(0) = 0.1$ and $\phi(w)$ then assumes the illustrative values given in Table 4.1.

Table 4.1

w	$\phi(w)$	w	$\phi(w)$
0	0.1000	60	0.7910
10	0.2608	70	0.8137
20	0.4109	80	0.8520
30	0.5321	90	0.8822
40	0.6283	100	0.9066
50	0.7047	110	0.9258

RUIN BEFORE "TIME" t WITH NO INITIAL RISK RESERVE

Since the following argument is quite general and yields a useful result apart from the determination of $\phi(0)$, let us write $u(0, t)$ for the probability that $\pi_1\tau$, the aggregate premium intake through operational time τ, is never less than the aggregate claim outgo for $0 < \tau \leq t$. Then $1 - u(0, t) \equiv v(0, t)$, the probability of ruin before epoch t, is a nondecreasing function of t bounded above by unity. In the limit, as t tends to infinity, we write

$$\phi(0) = \lim_{t \to \infty} u(0, t).$$

Later we consider the more general probability $u(x, t)$, namely that of not being ruined in the period $(0, t)$ when the initial risk reserve is $R(0) = x$.

Suppose that n claims have occurred up to time t and that the resulting aggregate claim is $Y(t) = y$. The joint probability of these events is

$$e^{-t} \frac{t^n}{n!} \, dP^{n*}(y).$$

Suppose that $\pi_1 t \geq y$; we then require the probability that the $n - 1$ increments in $Y(\tau)$ occurring at $n - 1$ different τ-values all less than t will not cause $Y(\tau)$ to exceed $\pi_1\tau$ for any $\tau \leq t$. This probability may be obtained from a theorem in analysis that is proved in Takács (1967, Chapter 1). We may paraphrase this theorem as follows:

$Y(\tau)$ is a nondecreasing step function with n steps in $(0, t)$ and $Y(0) = 0$. Define $Y(t + \tau) = Y(t) + Y(\tau)$, $0 \leq \tau \leq \infty$, and

$$\delta(\tau) = \begin{cases} 1, & \text{if } \pi_1\tau - Y(\tau) \leq \pi_1\sigma - Y(\sigma) \quad \text{for} \quad \tau \leq \sigma, \\ 0, & \text{otherwise.} \end{cases}$$

Then

$$\int_0^t \delta(\tau) \, d\tau = \pi_1 t - Y(t) \quad \text{provided} \quad Y(t) < \pi_1 t$$

The proof is achieved by means of a function $\chi(\cdot)$ defined by

$$\chi(\tau) \equiv \inf_{\sigma} [\pi_1\sigma - Y(\sigma)] \leq \tau - Y(\tau), \qquad \tau \leq \sigma,$$

so that

$$\chi(\sigma) - \chi(\tau) \leq \pi_1(\sigma - \tau),$$

implying monotonicity and continuity with

$$0 \leq \chi'(\tau) \leq \pi_1.$$

Furthermore

$$\chi(t + \tau) = \inf_{\sigma} [\pi_1(t + \sigma) - Y(t + \sigma)] = \inf_{\sigma} [\pi_1 t - Y(t) + \pi_1\sigma - Y(\sigma)]$$

$$= \pi_1 t - Y(t) + \inf_{\sigma} [\pi_1\sigma - Y(\sigma)] = \pi_1 t - Y(t) + \chi(\tau),$$

so that $\chi(t) = \pi_1 t - Y(t) + \chi(0)$. Hence

$$\int_0^t \chi'(\tau)\,d\tau = \chi(t) - \chi(0) = \pi_1 t - Y(t).$$

Noting that $\delta(\tau) = 1$ if and only if $\chi(\tau) = \pi_1\tau - Y(\tau)$, Takács proves that $\chi'(\tau) = \delta(\tau)$ for almost all τ. [We mention that if τ and $Y(\tau)$ only assume integer values we may define

$$\delta_r = \begin{cases} 1, & \text{if } \pi_1 r - Y(r) < \pi_1 s - Y(s) \quad \text{for} \quad r < s, \\ 0, & \text{otherwise,} \end{cases}$$

and the theorem states that $\sum_{r=1}^{t} \delta_r = t - Y(t)$]. It follows from the foregoing theorem that

$$\mathcal{I}\{Y(\tau) \leq \pi_1\tau \quad \text{for} \quad 0 \leq \tau \leq t \mid Y(t) = y\} = \frac{\displaystyle\int_0^t \delta(\tau)\,d\tau \quad \text{with} \quad Y(t) = y}{\displaystyle\int_0^t \delta(\tau)\,d\tau \quad \text{with} \quad Y(t) = 0}$$

$$= \frac{\pi_1 t - y}{\pi_1 t} = 1 - \frac{y}{\pi_1 t}.$$

The joint probability of n claims with an aggregate amount of y and no ruin before time t is thus

$$e^{-t} \frac{t^n}{n!} \left(1 - \frac{y}{\pi_1 t}\right) dP^{n*}(y), \qquad 0 \leq y \leq \pi_1 t,$$

and accordingly

$$u(0, t) = \sum_{n=0}^{\infty} e^{-t} \frac{t^n}{n!} \int_0^{\pi_1 t} \left(1 - \frac{y}{\pi_1 t}\right) dP^{n*}(y), \tag{4.8}$$

the term for zero claims being e^{-t}, since $P^{0*}(y) = \begin{cases} 0 & y < 0 \\ 1 & y \geq 0 \end{cases}$. This relation was first derived by Prabhu (1961).

It is obvious that if we had asked for the same probability, except for the stipulation that the risk reserve at time t should not be less than $z > 0$ (instead of 0), the only change in the result (4.8) would have been to replace the upper limit of the integral with $\pi_1 t - z.$*

We note that (4.8) can be written

$$u(0, t) = \sum_{n=0}^{\infty} e^{-t} \frac{t^n}{n!} (\pi_1 t)^{-1} \int_0^{\pi_1 t} P^{n*}(y)\, dy = (\pi_1 t)^{-1} \int_0^{\pi_1 t} F(y, t)\, dy$$

$$= (\pi_1 t)^{-1} \left\{ \int_{\pi_1 t}^{\infty} [1 - F(y, t)]\, dy - \int_0^{\infty} [1 - F(y, t)]\, dy + \pi_1 t \right\}$$

$$= 1 - \frac{p_1}{\pi_1} + \frac{p_{\pi_1 t}}{\pi_1 t}. \tag{4.9}$$

The numerical work on p_c in Bohman and Esscher (1963–1964, cited in Chapter 2) is thus relevant also to $u(0, t)$.

Illustration. As an illustration consider the case in which

$$P(y) = 1 - e^{-y}, \qquad 0 < y < \infty,$$

which, by (2.34), results in

$$P^{n*}(y) = 1 - e^{-y} \sum_{j=0}^{n-1} \frac{y^j}{j!}, \qquad n = 1, 2, 3, \ldots.$$

Relations 2.35 and 4.9 now give

$$1 - u(0, t) = (\pi_1 t)^{-1} \int_0^{\pi_1 t} e^{-y}\, dy \int_0^t e^{-s} I_0(2\sqrt{ys})\, ds$$

$$= \frac{1}{\pi_1 t} \sum_{j=0}^{\infty} \frac{1}{(j!)^2} \int_0^{\pi_1 t} y^j e^{-y}\, dy \int_0^t s^j e^{-s}\, ds$$

$$= \frac{1}{\pi_1 t} \sum_{j=0}^{\infty} \frac{\gamma(j + 1, \pi_1 t)\, \gamma(j + 1, t)}{\Gamma^2(j + 1)} \qquad \text{in gamma function notation}$$

$$= \frac{1}{\pi_1 t} \sum_{j=0}^{\infty} D(j + 1 \mid \pi_1 t)\, D(j + 1 \mid t), \tag{4.10}$$

* By writing $u(0, t \mid n, z)$ for the typical term in this expression and differentiating it a relation is obtained that can easily be verified for $n = 0$, and by making use of an expression for $d_z u(0, t \mid n, z)$ similar to that for $u(z, t)$ in (4.17) the general term can be proved inductively (Bhat, 1968).

where $D(k \mid \lambda) = e^{-\lambda} \sum_{j=k}^{\infty} \lambda^j/j!$. In the foregoing $I_0(\cdot)$ is the modified Bessel function we have already encountered in (2.35).

$$* \quad * \quad *$$

Let us now consider how to use the foregoing results to determine

$$\phi(0) = \lim_{t \to \infty} u(0, t).$$

There is no need to proceed to the limit in (4.8), for we can argue as follows:

$$\phi(0) = \mathfrak{I}\{Y(t) \leq \pi_1 t, \quad \text{all } t\} = \lim_{t \to \infty} \mathfrak{I}\{Y(\tau) \leq \pi_1 \tau, \quad \text{all } \tau \leq t\}$$

$$= \lim_{t \to \infty} \mathcal{E} \, \mathfrak{I}\{Y(\tau) \leq \pi_1 \tau, \quad \text{all } \tau \leq t \mid Y(t) = y \leq \pi_1 t\}$$

$$= \lim_{t \to \infty} \mathcal{E}\left\{\frac{\pi_1 t - y}{\pi_1 t}\right\} = \frac{\pi_1 - p_1}{\pi_1} = \frac{\eta_1}{p_1 + \eta_1}. \tag{4.11}$$

This is a general result valid for any $P(y)$ such that $P(0) = 0$. We note that $\psi(0) = 1 - \phi(0) = p_1/(p_1 + \eta_1)$ so that eventual ruin is certain unless $\eta_1 > 0$.

Using (4.11) and reverting to (4.7), we see that if all sums at risk are equal to h

$$\phi(w) = \frac{\pi_1 - h}{\pi_1} \sum_{j=0}^{[w/h]} \frac{(-1)^j}{j!} \left(\frac{w - hj}{\pi_1}\right)^j \exp\left(\frac{w - hj}{\pi_1}\right). \tag{4.12}$$

This is Erlang's (1909) formula when we write $\pi_1 = 1$.

THE DISCRETE-TIME CASE: RANDOM WALK

Suppose that in (4.3) the loaded premium π_1 is a unit payable at the beginning of each year during which at most one claim for a sum insured of h units can occur with probability q. Let w be the integer representing the risk reserve; then analogously to (4.4) we have

$$\phi(w) = \sum_{j=1}^{\infty} (1 - q)^{j-1} q \, \phi(w + j - h)$$

$$= q(1 - q)^{-w} \sum_{k=w}^{\infty} (1 - q)^k \phi(k + 1 - h), \tag{4.4'}$$

with $\phi(w) = 0$ for $w \leq 0$. [We have adopted the standard boundary condition of random walk.]

We introduce the generating function

$$G(s) = \sum_{w=0}^{\infty} s^w \, \phi(w).$$

Then by multiplying the foregoing equation by s^w and summing over w we have

$$G(s) = q \sum_{w=1}^{\infty} \left(\frac{s}{1-q}\right)^w \sum_{k=w}^{\infty} (1-q)^k \, \phi(k+1-h)$$

$$= q \sum_{k=1}^{\infty} (1-q)^k \, \phi(k+1-h) \sum_{w=1}^{k} \left(\frac{s}{1-q}\right)^w$$

$$= \frac{qs}{1-q-s} \sum_{k=1}^{\infty} [(1-q)^k - s^k] \, \phi(k+1-h)$$

$$= \frac{qs}{1-q-s} [(1-q)^{h-1} G(1-q) - s^{h-1} G(s)]$$

$$= (1-q-s+qs^h)^{-1} sq(1-q)^{h-1} G(1-q)$$

$$= (1-q-s+qs^h)^{-1} s(1-q) \, \phi(1), \qquad (4.5)'$$

since

$$\phi(1) = q \sum_{j=1}^{\infty} (1-q)^{j-1} \, \phi(j+1-h)$$

$$= q(1-q)^{h-2} \sum_{k=1}^{\infty} (1-q)^k \, \phi(k) = q(1-q)^{h-2} G(1-q).$$

This relation is analogous to (4.5). Consider the factors multiplying $\phi(1)$ in (4.5)', namely,

$$(1-q) \, s(1-q-s+qs^h)^{-1} = s\left[1 - \frac{s}{1-q}(1-qs^{h-1})\right]^{-1}$$

$$= s \sum_{j=0}^{\infty} \left(\frac{s}{1-q}\right)^j (1-qs^{h-1})^j$$

$$= s \sum_{j=0}^{\infty} (1-q)^{-j} s^j \sum_{k=0}^{j} (-1)^k \binom{j}{k} q^k s^{k(h-1)}$$

In the double sum last written a typical term is

$$(-1)^k \binom{j}{k} (1-q)^{-j} q^k s^{j+(h-1)k+1}.$$

We require the sum of all terms in which j and k satisfy the relation

$$j + (h-1)k + 1 = l, \qquad 0 \le k \le j,$$

For fixed k this relation is

$$j = l - 1 - (h-1)k$$

and the typical term can be written

$$(-1)^k \binom{l - 1 - \overline{h - 1}k}{k} [(1 - q)^{h-1}q]^k(1 - q)^{-l+1}s^l$$

for k, proceeding by integers from zero to the largest k satisfying

$$j = l - 1 - (h - 1)k \geq k, \qquad \text{i.e., } hk \leq l - 1.$$

Hence the required sum of terms multiplying s^l is

$$(1 - q)^{-l+1} \sum_{k=0}^{\overline{[l-1/h]}} (-1)^k \binom{l - 1 - \overline{h - 1}k}{k} [(1 - q)^{h-1}q]^k \equiv H(l),$$

$$l = 1, 2, \ldots,$$

which can be numerically evaluated.

Since $\phi(w)$ is the coefficient of s^w in $G(s)$, we have thus found that

$$\phi(w) = H(w) \phi(1), \tag{4.6'}$$

a result due to Burman (1946).

The analog of (4.8) assumes a simple form. We require $u(1, t)$, the probability that ruin will not occur within t years ($t = 1, 2, 3, \ldots$), given that the initial risk reserve is unity. Now, if j of the t years result in claims, the probability of which is $\binom{t}{j}q^j(1 - q)^{t-j}$, we require two conditions to be satisfied:

1. The aggregate premiums received must equal or exceed the aggregate claims paid, that is, $t \geq hj$.
2. The aggregate of premiums received at any year end during the period must be at least h times the claims that have occurred before that year end.

We determine the probability of the second event, given that the first holds true, as follows: write $L(j, t)$ for the multiple event in which the aggregate claims have always been less than the premiums and $E(j, t)$ for the event in which equality of claims and premiums can also occur. Now $1 - j/t$ is the probability that the first premium is not followed by a claim and we see that

$$\mathfrak{I}\{L(j, t)\} = \left(1 - \frac{j}{t}\right) \mathfrak{I}\{E(j, t - 1)\}$$

whereas $\mathfrak{I}\{L(j, t)\} = k - hj/t$ by Takác's theorem used in deriving (4.8). Thus

$$\mathfrak{I}\{E(j, t)\} = \frac{\mathfrak{I}\{L(j, t + 1)\}}{1 - j/(t + 1)} = \frac{t + 1 - hj}{t + 1 - j}$$

and

$$u(1, t) = \sum_{j=0}^{[t/h]} \binom{t}{j} q^j (1 - q)^{t-j} \frac{t + 1 - hj}{t + 1}$$

$$= (1 - q)^{-1} B\left(\left[\frac{t}{h}\right] \middle| t + 1, q\right) - hq(1 - q)^{-1} B\left(\left[\frac{t}{h}\right] - 1 \middle| t, q\right),$$

$$(4.8)'$$

where $B(x \mid t, q)$ represents the cumulation through the $(x + 1)$th term of the binomial with parameters t and q.

Finally, we may repeat the arguments leading to (4.11) to obtain

$$u(1, \infty) = \phi(1) = \lim_{t \to \infty} \mathcal{E}_j\left(\frac{t + 1 - hj}{t + 1 - j}\right) = 1 - \frac{(h - 1)q}{1 - q} = \frac{1 - hq}{1 - q} \quad (4.11)'$$

and from (4.6)'

$$\phi(w) = (1 - hq)(1 - q)^{-w} \sum_{k=0}^{\overline{[w-1/h]}} (-1)^k \binom{w - 1 - \overline{h - 1}k}{k} [(1 - q)^{h-1}q]^k$$

The foregoing special (discrete) case of (4.3) is an example of the random walk in one dimension. The name derives from the concept of a particle constrained to coincide with one of the integer values on the positive half of the X-axis. It is initially at the value w and moves one integer to the right if a certain event (no claim) occurs with probability $1 - q$. On the other hand, if the event does not occur, the particle moves $h - 1$ integers to the left. This procedure is repeated indefinitely unless the particle is absorbed by reaching the origin of the X-axis. It is convenient to provide that if the particle reaches the integer $M \gg w$ it stalls there until the event once more fails to occur when it moves $h - 1$ integers to the left. The barrier M may tend to infinity. The history of this random walk is given in Seal (1966).

Illustration.   Suppose that an employer who is paying the medical bills of his employees has found that these bills aggregate $100,000 a month on the average and that once in 10 months the aggregate exceeds $230,000, the average excess then being $100,000. He has decided to set up a risk reserve that will reimburse the company for $90,000 of its medical-bill outgo each time the aggregate exceeds $230,000. The risk reserve is to be set up with an initial *corpus* of 10^4 w and thereafter it will be credited with a monthly premium of $10,000.

The probability that the risk-reserve will never have to seek additional funds is given by $\phi(w)$, where $q = 0.1$ and $h = 9$. Illustrative numerical values are listed in Table 4.2. These values may be compared with the corresponding continuous-time values following relation (4.7).

Table 4.2

w	$\phi(w)$	w	$\phi(w)$
1	0.1111	60	0.7996
10	0.2745	70	0.8451
20	0.4376	80	0.8803
30	0.5656	90	0.9076
40	0.6643	100	0.9285
50	0.7406	110	0.9449

*   *   *

It is of interest to use standard partial-fraction techniques on (4.5)′ to obtain asymptotic results for $\phi(w)$. Write

$$G(s) = \frac{s(1 - hq)}{1 - q - s + qs^h} \equiv \frac{U(s)}{V(s)} \equiv \sum_{j=1}^{h} \frac{a_j}{s_j - s},$$

where s_j $(j = 1, 2, 3, \ldots, h)$ are the roots of $V(s) = 0$ written in order of increasing modulus and (Feller, 1968, Chapter XI, cited in Chapter 3)

$$a_j = \frac{-U(s_j)}{V'(s_j)} = \frac{s_j(1 - hq)}{1 - hqs_j^{h-1}} = \frac{s_j^2(1 - hq)}{(1 - q)h - (h - 1)s_j}.$$

Then the coefficient of s^w $(w = 1, 2, 3, \ldots)$ in $G(s)$ is

$$\phi(w) = \sum_{j=1}^{h} \frac{a_j}{s_j w + 1} \sim \frac{a_1}{s_1 w + 1} + \frac{a_2}{s_2^{w+1}}.$$

Consider the polynomial equation in s (h being integral):

$$qs^h - s + (1 - q) = 0.$$

It has h roots, of which, by Descartes' rule of signs, at most two are real and positive, and it can be verified that there are exactly two such roots, one being $s_1 = 1$ and the other, $s_2 \equiv s_0 > 1$. There is a single real negative root if h is odd and its modulus, and the moduli of all the imaginary roots, is greater than $|s_0|$. Hence

$$\phi(w) \sim 1 - \frac{1 - hq}{(h - 1)s_0 - h(1 - q)} s_0 - w + 1,$$

where s_0 is the larger of the two positive roots of the equation

$$qs^h - s + (1 - q) = 0;$$

for example, with $q = 0.1$ and $h = 9$,

$$s_0 \simeq 1.026119 = (0.9745458)^{-1} = (0.7727210)^{-1/10},$$
$$\phi(w) \sim 1 - 0.941808(0.772721)^{w/10},$$

and we obtain the specimen values of $\phi(w)$ given in Table 4.3. These values

Table 4.3

w	$\phi(w)$	w	$\phi(w)$
1	0.0822	60	0.7995
10	0.2722	70	0.8451
20	0.4376	80	0.8803
30	0.5655	90	0.9075
40	0.6642	100	0.9285
50	0.7405	110	0.9448

agree remarkably well with the exact values for $w \geq 20$.

*   *   *

Seal (1966) considered $[u(x, t - 1) - u(x, t)]/[1 - u(x, \infty)]$, namely, the probability that ruin occurs at the tth step, given that it occurs eventually, and obtained asymptotic values for the first few moments of the random variable T to show numerically that the third and fourth moments were close to the Normal values for small q. Exact formulas for $u(x, t - 1) - u(x, t)$ in the special case in which $h = 2$ are given in Fürst (1956) and Weesakul (1961). The more general random walk in which the size of any leftward step is determined by a probability distribution is considered by Klinger (1965).

THE PROBABILITY OF RUIN BEFORE "TIME" t

Write $u(z, t) = \mathcal{P}\{R(\tau)$ is never ≤ 0 for $0 \leq \tau \leq t$, given that the initial risk reserve is $z\}$, $R(0) = z$. If $v(z, t) = 1 - u(z, t)$, we notice that $v(z, t)$ is a nondecreasing function of t bounded above by unity.

Suppose that the first claim occurring at time τ amounts to y, with probability $dP(y)$. The probability of not being ruined in the following period of time $t - \tau$ is then $u(z + \pi_1\tau - y, t - \tau)$.

We thus have

$$u(z, t) = \int_0^\infty e^{-\tau}\, d\tau \int_0^{z+\pi_1\tau} u(z + \pi_1\tau - y, t - \tau)\, dP(y), \qquad p_1 + \eta_1 \equiv \pi_1.$$

$$(4.13)$$

By writing $z + \pi_1\tau = s$ this becomes

$$u(z, t) = \pi_1^{-1} \int_z^\infty e^{-(s-z)/\pi_1}\, ds \int_0^s u\left(s - y, t - \frac{s - z}{\pi_1}\right) dP(y),$$

and on differentiating with respect to z (Whittaker and Watson, 1927)

$$\pi_1 \frac{\partial u(z, t)}{\partial z} = \int_z^\infty \frac{\partial}{\partial z}\left[e^{-(s-z)/\pi_1} \int_0^s u\left(s - y, t - \frac{s - z}{\pi_1}\right) dP(y)\right] ds$$

$$- e^{-(s-z)/\pi_1} \int_0^s u\left(s - y, t - \frac{s - z}{\pi_1}\right) dP(y)\bigg|_{s=z}$$

$$= \pi_1^{-1} \int_z^\infty e^{-(s-z)/\pi_1}\, ds \int_0^s \left[\frac{\partial}{\partial t} u\left(s - y, t - \frac{s - z}{\pi_1}\right)\right.$$

$$\left. + u\left(s - y, t - \frac{s - z}{\pi_1}\right)\right] dP(y) - \int_0^z u(z - y, t)\, dP(y)$$

$$= \frac{\partial}{\partial t} u(z, t) + u(z, t) - \int_0^z u(z - y, t)\, dP(y). \tag{4.14}$$

This equation, a form of the Fokker-Planck equation (Feller, 1966), was derived by Arfwedson (1950), and later (with $\pi_1 = 1$) by Takács (1955) where $u(z, t)$ represented the probability that a customer arriving at the tail of an orderly queue at time t after an epoch when the server was idle would have a wait of at most z before he was served (namely, when the customer just ahead of him had his service completed). The assumption was that customers arrived at random with an average interarrival time of unity (just as the average period between two claims is unity). As each customer joined the queue the waiting-time jumped upward by a random amount y with d.f. $P(y)$, and between customer arrivals the waiting-time curve was a straight line downward with slope -1 unless it reached the t-axis, after which the waiting time remained zero until a new customer arrived. Let $W(t)$ be the waiting time for a customer joining the queue at time t, and let $\overline{W}(t)$ be constructed in the same way as $W(t)$, except that negative values are allowed so that the next following upward jump will occur from a negative instead of a zero value. It can then be shown that

$$\mathcal{I}\{W(t) \leq z\} = \mathcal{I}\left\{\sup_{0 < \tau \leq t} \overline{W}(\tau) \leq z\right\},$$

and this is the same as the probability, in our case, that the reduction in the initial risk reserve will not exceed z at some point during the interval $(0, t)$.

An earlier version of (4.14), appropriate for an annuity fund (*post*), was obtained by Saxén (1948), and for the special case in which $\pi_1 = 0$ and $u(x, t)$ becomes $F(x, t)$ the equation appears in Lundberg (1909).

Equation 4.14 may be written in the alternate form

$$\pi_1 \frac{\partial u(z,t)}{\partial z} - \frac{\partial u(z,t)}{\partial t} = \frac{\partial}{\partial z} \int_0^z u(z-y,t)[1-P(y)]\, dy,$$

and, on putting $1 - P(y) = Q(y)$ and $\pi_1^{-1} Q(y) = h(y)$,

$$\frac{\partial}{\partial z}\left[u(z,t) - \int_0^z u(z-y,t)\, h(y)\, dy \right] = \pi_1^{-1} \frac{\partial}{\partial t} u(z,t).$$

Integrating this with respect to z between 0 and x, we obtain

$$u(x,t) - u(0,t) - \int_0^x u(x-y,t)\, h(y)\, dy = \pi_1^{-1} \frac{\partial}{\partial t} \int_0^x u(z,t)\, dz. \quad (4.15)$$

When $t = \infty$, the case in which ruin never occurs, the right-hand term is zero and the equation for $u(x, \infty)$ is a special case of the "renewal equation" discussed below.

The general integro-differential equation (4.15) cannot be solved explicitly, and numerical results must be obtained by approximate methods. Because of the rapidity with which $u(x, t)$ moves away from its boundary value $u(x, 0) = 1$ as t increases through a set of points $j\delta$ ($j = 0, 1, 2, \ldots ; \delta > 0$), approximate integration formulas would be laborious to apply, even on a fast computer. The asymptotic relations derived for the difference $u(x, t) - u(x, \infty)$ (Cramér, 1955; Pollaczek, 1957) do not produce useful numerical estimates for $u(x, t)$. On the other hand, computer simulation of relation (4.3) is quite straightforward, and it is, perhaps, surprising that there have been so few attempts to calculate $u(x, t)$ in the literatures of queueing or risk theory.

A possibly more convenient equation for $u(x, t)$ can be derived from the following considerations: $F(z + \pi_1 t, t)$ is the probability that the aggregate outgo during time t will not exceed $z + \pi_1 t$, so that the business is in a state of nonruin at time t. During this time, however, "ruin," or rather a negative risk reserve, could have occurred on several occasions. (We assume that the company was able to borrow funds to pay claims when its reserves were negative. These borrowings were to be paid from subsequent premiums.) If ruin did not occur, $u(z, t)$ is equivalent to the probability that the outgo has not exceeded $z + \pi_1 t$. Suppose, on the other hand, ruin occurred at least once and that after the *last* time this happened the upward sloping line $R(t)$ crossed the t-axis at time $\tau < t$. The probability that the aggregate outgo up to time τ amounted to exactly $z + \pi_1 \tau$ will be written as

$$d_y F(y, \tau)\big|_{y=z+\pi_1\tau} \equiv dF(z + \pi_1 \tau, \tau), \qquad \tau > 0,$$

and, since this implies a zero risk reserve at that time, the probability of no ruin between time τ and time t is $u(0, t - \tau)$. We thus obtain (Beneš, 1960; Prabhu, 1961)

$$F(z + \pi_1 t, t) = u(z, t) + \int_{\tau=0+}^{t} u(0, t - \tau) \, dF(z + \pi_1 \tau, \tau), \qquad (4.16)$$

where $F(y, t)$ is given by (2.33) and z and $\pi_1 = p_1 + \eta_1$ are known parameters.

Notice that $u(0, t)$ occurs in both (4.15) and (4.16). We have already derived an expression for it in (4.8).

Illustration. Let us illustrate the application of relation (4.16) to the particular case of equal sums at risk. We have

$$P(y) = P^{1*}(y) = \begin{cases} 0, & y < h, \\ 1, & y \geq h, \end{cases}$$

and from (2.38)

$$P^{n*}(y) = \overline{P(y - n - 1h)}. \qquad (4.17)$$

Thus

$$\int_0^{\pi_1 t} \left(1 - \frac{y}{\pi_1 t}\right) dP^{n*}(y) = \begin{cases} 1 - \dfrac{nh}{\pi_1 t}, & nh \leq \pi_1 t, \\ 0, & nh > \pi_1 t, \end{cases}$$

and from (4.8),

$$u(0, t) = \sum_{n=0}^{[\pi_1 t/h]} e^{-t} \frac{t^n}{n!} \left(1 - \frac{nh}{\pi_1 t}\right). \qquad (4.18)$$

Furthermore, from (2.33)

$$F(y, t) = \sum_{j=0}^{\infty} e^{-t} \frac{t^j}{j!} \, \overline{P(y - j - 1h)} = \sum_{j=0}^{[y/h]} e^{-t} \frac{t^j}{j!} \qquad (4.19)$$

and

$$d_y F(y, t) = e^{-t} \frac{t^{[y/h]}}{([y/h])!} \quad \text{when} \quad \left[\frac{y}{h}\right] = \frac{y}{h} \text{ and is zero otherwise.}$$

Consider the Poisson probabilities assumed by $d_y F(y, t)$ when $y = z + \pi_1 \tau$ and τ increases from $\varepsilon > 0$ to t. We are interested only in integer values of $[y/h] \equiv n$, say, and we write n_1 for the smallest value assumed by n and n_2 for its largest value. In general

$$n = \left[\frac{y}{h}\right] = \left[\frac{z + \pi_1 \tau}{h}\right]$$

and, when $\tau = \varepsilon$ arbitrarily small,

$$n_1 = \left[\frac{z}{h}\right] + 1.$$

On the other hand, the largest value of n is

$$n_2 = \left[\frac{z + \pi_1 t}{h} \right].$$

For intermediate n-values τ is given by

$$\tau = \frac{nh - z}{\pi_1},$$

which, in general, is not integral.

Hence, using (4.16),

$$u(z, t) = \sum_{j=0}^{[(z+\pi_1 t)/h]} e^{-t} \frac{t^j}{j!} - \sum_{n=[z/h]+1}^{[(z+\pi_1 t)/h]} e^{-(nh-z)/\pi_1} \frac{[(nh - z)/\pi_1]^n}{n!} u\left(0, t - \frac{nh - z}{\pi_1} \right),$$

(4.20)

where $u[0, t - (nh - z)/\pi_1]$ may be obtained from (4.18). Note that when $t \to \infty$

$$u(z, \infty) \equiv \phi(z) = 1 - \phi(0) \sum_{j=[z/h]+1}^{\infty} \frac{1}{j!} e^{(hj-z)/\pi_1} \left(\frac{hj - z}{\pi_1} \right)^j,$$

which is equivalent to (4.12).

An explicit result for $u(z, t)$ can also be obtained when $P(y) = 1 - e^{-y}$ (Arfwedson, 1950; Prabhu, 1961).

GENERALIZATIONS OF POISSON ASSUMPTION

Let us rewrite (4.8) as follows:

$$u(0, t) = \int_0^{\pi_1 t} \left(1 - \frac{y}{\pi_1 t} \right) \sum_{n=0}^{\infty} e^{-t} \frac{t^n}{n!} \, dP^{n*}(y) = \int_0^{\pi_1 t} \left(1 - \frac{y}{\pi_1 t} \right) d_y F(y, t).$$

It is clear that we may generalize the Poisson assumption for $p_n(t)$, provided $(1 - y/\pi_1 t)$, which does not involve n, remains the probability that the income $\pi_1 t$ has always exceeded the claim outgo throughout the period $0 < \tau < t$; and once the generalized $p_n(t)$ has been used to determine $u(0, t)$ the appropriate value of $u(x, t)$ may be calculated from (4.16).

The most general formulation of the validity of the foregoing relation for $u(0, t)$ appears to be that of Takács (1967). The stochastic process $Y(t)$ which appears in relation (4.3) is said to have cyclically interchangeable increments if, for all n, the n cyclic permutations* of the random variables

$$\Delta_j = Y\left(\frac{j}{n} t \right) - Y\left(\frac{j - 1}{n} t \right), \qquad j = 1, 2, 3, \ldots, n,$$

* The kth cyclic permutation is $\Delta_k, \Delta_{k+1}, \Delta_{k+2}, \ldots, \Delta_n, \Delta_1, \Delta_2 \cdots \Delta_{k-1}$ $(k = 1, 2, \ldots, n)$.

all have the same joint distribution. Cyclic interchangeability includes ordinary interchangeability and also mutual independence with identical distribution of the n variables ($n = 1, 2, 3, \ldots$). Assuming cyclic interchangeability, the conditional probability

$$\mathscr{S}\{Y(\tau) \leq \pi_1 \tau \quad \text{for} \quad 0 \leq \tau \leq t \mid Y(t) = y\} = 1 - \frac{y}{\pi_1 t}.$$

If we require all claims to be independent and identically distributed, the only restriction on $p_n(t)$ is that imposed by (4.3), namely, that after suitable transformation of the time scale the distribution of the number of claims in any "time" interval depends only on the length of the interval. This is a property of stationary point processes (Cox and Lewis, 1966), which include the following:

1. The mixed Poisson process defined by (2.28) which is equivalent to a general birth process with stationary increments (McFadden, 1965).

2. The renewal process, namely, that in which interclaim periods are independently and identically distributed with density $f(t)$. When $f(t)$ is negative exponential, the renewal process becomes a Poisson process and (2.15) holds. On the other hand, a renewal process is not a mixed Poisson unless there is only one component in the mixing (McFadden, 1965).

3. The doubly stochastic Poisson process in which claims occur as the realization of a Poisson process in which $\lambda(\tau)$, the force of claim occurrence (or hazard) at time τ, is itself a stationary stochastic process.

4. The special case of (3) in which the length of an interclaim period depends only on the length of the preceding period (Wold, 1949).

5. The semi-Markov process in which there are interclaim periods of k different types, each having its own probability distribution of period lengths, the successive occurrences of different types being governed by a $k \times k$ Markov chain matrix of transition probabilities.

6. The branching renewal process in which interclaim periods are generated by the superposition of a renewal process of primary events, each of which generates a renewal process of subsidiary events.

The only actuarial applications of these more general possibilities appear to be Andersen's (1957) use of a $\lambda(\tau)$ which is a function of the length of the interval elapsed since the claim prior to time τ, and the mixed Poisson process referred to on a number of occasions in this monograph. Because of their likely importance in future attempts to apply more general models to a risk business, we have summarized some of the mathematics of renewal processes in Appendix A.

However, when $P(\cdot)$ is a function of τ, we must fall back on the cyclic interchangeability of the increments; for example, in the case in which $Y(\tau) = Y e^{-(t-\tau)\delta}$, with Y a non-negative random variable independent of τ, and $\delta > 0$ (namely, where interest is supposed to accrue on the risk reserve at an instanteous rate δ), although any increment depends on τ the whole set of n increments has a joint distribution that is independent of their order. Hence relations (4.8) and (4.16) apply, provided $P(\cdot)$ is replaced by

$$\frac{1}{t}\int_0^t P(y\overline{e^{t-\tau\delta}})\,d\tau.$$

THE LAPLACE TRANSFORM OF $u(z, t)$

We shall use the following notation hereafter:

$$\pi(s) = \int_0^\infty e^{-sy}\,dP(y) = s\int_0^\infty e^{-sy}\,P(y)\,dy.$$

[To distinguish these two integrals we call $\pi(s)$ the Laplace-Stieltjes Transform (L.S.T.) of $P(y)$, whereas the integral last written, which equals $\pi(s)/s$, is the L.T. of $P(y)$.]

$$h(y) = \frac{1 - P(y)}{\pi_1},$$

$$\begin{aligned}
\chi(s) &= \int_0^\infty e^{-sy}\,h(y)\,dy = \pi_1^{-1}\int_0^\infty e^{-sy}\,dy\int_y^\infty dP(z)\\
&= \pi_1^{-1}\int_0^\infty dP(z)\int_0^z e^{-sy}\,dy = (s\pi_1)^{-1}\int_0^\infty (1 - e^{-sz})\,dP(z)\\
&= (s\pi_1)^{-1}[1 - \pi(s)], \tag{4.21}
\end{aligned}$$

and

$$\mu(r, s) = \int_0^\infty e^{-rx}\,dx\int_0^\infty e^{-st}\,u(x, t)\,dt. \tag{4.22}$$

It is easily verified that

$$[\pi(s)]^n = \int_0^\infty e^{-sy}\,d\,P^{n*}(y), \qquad n = 1, 2, 3, \ldots.$$

As a preliminary to taking the bivariable L.T. of both sides of (4.15), we evaluate

$$\begin{aligned}
\int_0^\infty e^{-rx}\,dx\int_0^x u(x - y, t)\,h(y)\,dy &= \int_0^\infty h(y)\,dy\int_y^\infty e^{-rx}\,u(x - y, t)\,dx\\
&= \int_0^\infty e^{-ry}\,h(y)\,dy\int_y^\infty e^{-r(x-y)}\,u(x - y, t)\,dx\\
&= \chi(r)\int_0^\infty e^{-rz}\,u(z, t)\,dz,
\end{aligned}$$

$$\int_0^\infty e^{-rx}\,dx\int_0^x u(z, t)\,dz = \int_0^\infty u(z, t)\,dz\int_z^\infty e^{-rx}\,dx = \frac{1}{r}\int_0^\infty e^{-rz}\,u(z, t)\,dz,$$

and

$$\int_0^\infty e^{-st}\, dt\, \frac{\partial}{\partial t} \int_0^\infty e^{-rz}\, u(z,t)\, dz = \int_0^\infty e^{-rz}\, dz \int_0^\infty e^{-st}\, \frac{du(z,t)}{dt}\, dt$$

$$= \int_0^\infty e^{-rz}\, dz \left[e^{-st} u(z,t) \Big|_{t=0}^\infty + s \int_0^\infty e^{-st} u(z,t)\, dt \right]$$

$$= s\, \mu(r,s) - \int_0^\infty e^{-rz}\, u(z,0)\, dz$$

$$= s\, \mu(r,s) - \frac{1}{r}, \quad \text{since} \quad u(z,0) = 1.$$

The bivariable L.T. of (4.15) is thus

$$\mu(r,s) - \frac{1}{r} \int_0^\infty e^{-st} u(0,t)\, dt - \chi(r)\, \mu(r,s) = \pi_1^{-1} \frac{1}{r}\left[s\, \mu(r,s) - \frac{1}{r} \right]$$

or, making use of the relation for $\chi(r)$ and writing $\mu(s) = \int_0^\infty e^{-st} u(0,t)\, dt$,

$$r[\pi_1 r - 1 - s + \pi(r)]\, \mu(r,s) - \pi_1 r\, \mu(s) + 1 = 0. \tag{4.23}$$

If $\rho(s)$ is that value of r that satisfies Lundberg's (1928) "fundamental" equation, namely,

$$s = \pi_1 r + \pi(r) - 1, \tag{4.24}$$

the term involving $\mu(r,s)$ in (4.23) vanishes when $r = \rho(s)$ and

$$\mu(s) = \frac{1}{\pi_1 \rho(s)}. \tag{4.25}$$

Then (4.23) provides

$$\mu(r,s) = \frac{1/r - 1/\rho(s)}{s + 1 - \pi_1 r - \pi(r)}. \tag{4.26}$$

These relations are due to Arfwedson (1954) who studied (4.24) in detail. They are, however, implicit in Segerdahl's (1939) work on the moments of the time to ruin. Essentially, the same formulas were derived for queueing theory by Beneš (1957). We are not aware of any numerical inversions of (4.26) for given $T(\cdot)$ and it would seem that such a procedure would be an awkward way of calculating $u(x,t)$. Nevertheless, because of the relative novelty of numerical methods for inverting Laplace transforms, we have reviewed the rather scattered literature in Appendix B. It will be seen that all four techniques hitherto suggested are quite simple mathematically.

ASYMPTOTIC VALUES OF THE MOMENTS OF THE TIME TO RUIN

Relation 4.26 can be used to derive an expression for the kth moment of the conditional random variable: time to ruin, given that ruin is certain.

We write this moment as $[1 - \phi(x)]^{-1} a_k(x)$, where $\phi(x)$ is the complement of the probability of ultimate ruin with a risk reserve of x. Then

$$a_k(x) = \int_0^\infty t^k \, d_t v(x, t) = -\int_0^\infty t^k \, d_t u(x, t),$$

with the result that $a_0(x) = u(x, 0) - u(x, \infty) = 1 - \phi(x)$, as it should. These moments may not be finite.

We follow the method adumbrated by Segerdahl (1955) to derive asymptotic values of the first two moments. A theorem in Laplace transforms (e.g., Thomson, 1960) states, in effect, that, provided $\lim_{r \to \infty} |\mu(r, s)| = 0$, the inverse (with respect to r) of the transform $\mu(r, s)$ is

$$\int_0^\infty e^{-st} u(x, t) \, dt$$

$$= \frac{1}{2\pi i} \int_{\gamma - i\infty}^{\gamma + i\infty} e^{rx} \mu(r, s) \, dr \qquad (4.27)$$

$$= \text{sum of the residues with respect to } r \text{ of } \quad \frac{1 - r/\rho(s)}{r[s + 1 - \pi_1 r - \pi(r)]} \, e^{rx}. \qquad (4.28)$$

Now the function

$$f(r) \equiv s + 1 - \pi_1 r - \pi(r) = s + 1 - \pi_1 r - \int_0^\infty e^{-ry} \, dP(y) \qquad (4.29)$$

has only two zeros on the positive real axis, namely, at $r = \rho$ and $r = R > \rho$, say. Other zeros are of absolute value greater than R. An asymptotic relation for (4.27) is thus obtained as the sum of the three residues $r_0(s)$ at the simple pole $r = 0$, $r_1(s)$ at the simple pole $r = \rho$, and $r_2(s)$ at the simple pole $r = R$; but

$$r_0(s) = \lim_{r \to 0} (r - 0) e^{rx} \mu(r, s) = \frac{1}{s},$$

$$r_1(s) = \lim_{r \to \rho} (r - \rho) e^{rx} \mu(r, s) = \lim_{r \to \rho} \frac{-[(r - \rho)^2 x + 2(r - \rho)] e^{rx}/\rho}{[s + 1 - \pi_1 r - \pi(r)] - r[\pi_1 + \pi'(r)]}$$
$$\text{by L'Hôpital's rule}$$

$$= 0,$$

and

$$r_2(s) = \lim_{r \to R} (r - R) e^{rx} \mu(r, s) = \lim_{r \to R} \frac{-[(r - R)(r - \rho)x + (2r - R - \rho)] e^{rx}/\rho}{[s + 1 - \pi_1 r - \pi(r)] - r[\pi_1 + \pi'(r)]}$$

$$= \frac{(R - \rho) e^{Rx}}{\rho R[\pi_1 + \pi'(R)]}.$$

Hence by (4.28) for large values of x

$$\int_0^\infty e^{-st} u(x, t) \, dt = \frac{1}{s} + \frac{1}{s} \int_0^\infty e^{-st} \, d_t u(x, t) \sim \frac{1}{s} + \frac{(R - \rho) e^{Rx}}{\rho R[\pi_1 + \pi'(R)]}$$

or

$$-\int_0^\infty e^{-st}\, d_t u(x,t) = \sum_{k=0}^\infty \frac{(-s)^k}{k!}\, a_k(x) \sim \frac{s(\rho - R)e^{Rx}}{\rho\, R[\pi_1 + \pi'(R)]}.$$ (4.30)

In order to obtain $a_k(x)$, $k = 0, 1, 2$, we thus require the first and second derivatives of the right-hand member of (4.30) with respect to s and must evaluate these derivatives and the function itself when $s = 0$.

Let us first consider (4.29) when r assumes the values $\rho \equiv \rho(s)$ and $R \equiv R(s)$, respectively, the two real positive roots of the equation $f(r) = 0$. When $s = 0$, $\rho(0) = \rho_0$ is determined from

$$1 - \pi_1 \rho_0 - \int_0^\infty e^{-\rho_0 y}\, dP(y) = 0,$$

so that $\rho_0 = 0$. Differentiation of the equation $f(r) = 0$ with respect to s when $r = \rho(s)$ produces

$$1 - \pi_1 \rho' + \rho' \int_0^\infty y e^{-\rho y}\, dP(y) = 0$$

and, differentiating again,

$$-\pi_1 \rho'' - \rho'^2 \int_0^\infty y^2 e^{-\rho y}\, dP(y) + \rho'' \int_0^\infty y e^{-\rho y}\, dP(y) = 0.$$

Differentiating a third time

$$-\pi_1 \rho''' + \rho'^3 \int_0^\infty y^3 e^{-\rho y}\, dP(y)$$

$$- 3\rho'\rho'' \int_0^\infty y^2 e^{-\rho y}\, dP(y) + \rho''' \int_0^\infty y e^{-\rho y}\, dP(y) = 0.$$

On putting $s = 0$, these expressions become, respectively,

$$\rho'(0) \equiv \rho_0' = \frac{1}{\pi_1 - \rho_1} = \frac{1}{\eta_1} \quad \rho_0'' = -\frac{p_2}{\eta_1{}^3} \quad \text{and} \quad \rho_0''' = \frac{\eta_1 p_3 + p_2{}^2}{\eta_1{}^5}.$$ (4.31)

In the case in which $r = R(s)$ the value of $R(0) \equiv R_0$ is found from

$$1 - \pi_1 R_0 - \int_0^\infty e^{-R_0 y}\, dP(y) = 0.$$ (4.32)

Writing

$$p_j' = \int_0^\infty e^{-R_0 y} y^j\, dP(y), \quad j = 0, 1, 2, 3,$$

and $\pi_1 - p_1' = \eta_1'$ the three relations (4.31) can be extended to this case, namely,

$$R'(0) \equiv R_0' = \frac{1}{\eta_1'}, \quad R_0'' = -\frac{p_2'}{\eta_1'^3} \quad \text{and} \quad R_0''' = \frac{\eta_1' p_3' + 3p_2'^2}{\eta_1'^5}.$$ (4.33)

We may now consider the expression on the right of (4.30), which we write as $g(s)$. Then

$$\ln g(s) = (\ln s - \ln \rho) + \ln (\rho - R) + Rx + \ln R' - \ln R,$$

$$\frac{g'(s)}{g(s)} = \left(\frac{1}{s} - \frac{\rho'}{\rho}\right) + \frac{\rho' - R'}{\rho - R} + R'x + \frac{R''}{R'} - \frac{R'}{R},$$

and

$$\frac{g''(s)}{g(s)} - \left[\frac{g'(s)}{g(s)}\right]^2 = \left[-\frac{1}{s^2} - \frac{\rho''}{\rho} + \left(\frac{\rho'}{\rho}\right)^2\right] + \frac{\rho'' - R''}{\rho - R}$$

$$- \left(\frac{\rho' - R'}{\rho - R}\right)^2 + R''x + \frac{R'''}{R'} - \left(\frac{R''}{R'}\right)^2 - \frac{R''}{R} + \left(\frac{R'}{R}\right)^2.$$

On putting $s = 0$ into these expressions (utilizing L'Hôpital's rule on the first terms) and substituting from (4.31), (4.32), and (4.33), we obtain

$$a_0(x) \sim g(0) = -\frac{\eta_1}{\eta_1'} e^{R_0 x}, \qquad (\eta_1' < 0),$$

$$\frac{a_1(x)}{a_0(x)} \sim -\frac{g'(0)}{g(0)} = -\frac{p_2}{2\eta_1^2} + \frac{1}{R_0\eta_1} - \frac{x}{\eta_1'} + \frac{p_2'}{\eta_1'^2}, \qquad (4.34)$$

$$\frac{a_2(x)}{a_0(x)} - \left[\frac{a_1(x)}{a_0(x)}\right]^2 \sim -\frac{1}{3} \cdot \frac{p_3}{\eta_1^3} - \frac{3}{4} \cdot \frac{p_2^2}{\eta_1^4} + \frac{p_2}{R_0\eta_1^3}$$

$$- \frac{1}{R_0^2\eta_1^2} + \frac{2}{R_0^2\eta_1\eta_1'} - \frac{p_2'x}{\eta_1'^3} + \frac{p_3'}{\eta_1'^3} + \frac{2p_2'^2}{\eta_1'^4}. \quad (4.35)$$

Relations 4.34 and 4.35 are the asymptotic values of the mean and variance of the distribution of the time to ruin, given that ruin is certain. By transferring the origin to the mean and using the square root of the variance as the time unit, we find (Segerdahl, 1955) that (4.30) can be written in the form

$$\exp\left[\frac{s^2}{2} + \sum_{j=3}^{\infty} \frac{b_j(x)}{x^{j/2}} s^j\right],$$

where the $b_j(x)$ are polynomials in x^{-1} with a constant term. Thus the distribution of the time to ruin, given that it is certain, is asymptotically Normal for large values of x. Since the mean (4.34) is a linear function of x, the approximation will hold good only for large values of t.

Illustration. When

$$P(y) = 1 - e^{-y}, \qquad 0 \le y < \infty,$$

it is easily verified that

$$p_1 = 1 \qquad p_2 = 2, \qquad p_3 = 6,$$
$$p_1' = (1+\eta)^2, \qquad p_2' = 2(1+\eta)^3, \qquad p_3' = 6(1+\eta)^4,$$
$$R_0 = \frac{\eta}{1+\eta}, \qquad \eta_1' = -\eta(1+\eta), \quad \text{and} \quad -\frac{\eta_1}{\eta_1'} = \frac{1}{1+\eta}.$$

The mean and variance of the asymptotically Normal distribution of the conditional distribution of the time to ruin are thus

$$\mu = -\frac{1}{\eta^2} - \frac{1+\eta}{\eta^2} + \frac{x}{\eta(1+\eta)} + \frac{2(1+\eta)}{\eta^2} = \frac{1}{\eta} + \frac{x}{\eta(1+\eta)}$$

and

$$\sigma^2 = -\frac{2}{\eta^3} - \frac{3}{\eta^4} - \frac{2(1+\eta)}{\eta^4} - \frac{(1+\eta)^2}{\eta^4} - \frac{2(1+\eta)}{\eta^4}$$
$$+ \frac{2x}{\eta^3} - \frac{6(1+\eta)}{\eta^3} + \frac{8(1+\eta)^2}{\eta^4}$$
$$= \frac{2+\eta}{\eta^3} + \frac{2x}{\eta^3}.$$

Hence the probability of ruin within a period during which expected deaths number t is

$$1 - u(x, t) \sim \frac{1}{1+\eta} e^{-\eta x/(1+\eta)} \Phi \left\{ \frac{\left[t - \dfrac{1}{\eta} - \dfrac{x}{\eta(1+\eta)}\right]}{\left(\dfrac{2+\eta}{\eta^3} + \dfrac{2x}{\eta^3}\right)^{1/2}} \right\}$$

where $\Phi(\cdot)$ is now the Normal d.f. with zero mean and unit variance.

Suppose, for example, that a company with a risk reserve of 50 times the mean claim and a risk loading of 5% of the premium decides that its claim-size distribution is near enough to the negative exponential for it to use the foregoing results. Its probability of eventual ruin is then

$$\frac{1}{1+\eta} e^{-\eta x/(1+\eta)} = \frac{1}{1.05} e^{-2.38095} = 0.08806.$$

This is relatively large, and the question is whether the probability of ruin during a time interval when expected claims number 100 is much smaller. The answer is given approximately by

$$1 - u(50, 100) \sim 0.08806 \times \Phi\left(\frac{100 - 20 - 952.381}{(102.05/0.000125)^{1/2}}\right)$$
$$= 0.08806 \times \Phi(-0.9655) = 0.08806 \times 0.16715$$
$$= 0.015.$$

The managers of the risk business would have to decide whether this "short run" probability of ruin was small enough to be accepted. If not, the capital, the premium loading, or both would have to be increased.

THE ANNUITY OR PENSION FUND

In the foregoing we have considered a claim distribution that is zero for negative Y. The case of purely negative Y-values can be made to correspond to an annuity business.

Consider a portfolio of immediate life annuities whose single-sum values at any given point of time constitute a random sample of variates Y with d.f. $B(\cdot)$. The annuity company pays annuities to its policyholders at the rate $p_1 - \eta_1$ per expected death per time interval, where p_1 is the mean of Y and η_1 is a risk loading in favor of the company. On the death of any annuitant the company is relieved of future annuity payments to that individual. Ignoring interest and business expense loadings, we may suppose that a separate department of the annuity company accepts the initial net single premiums which are equal to the expected values of future annuity payments to any specified annuitant, the distribution of these premiums being that of Y. The risk reserve makes the annuity payments continuously at the over-all average rate of $p_1 - \eta_1$. When the annuitant dies, the annuity department pays that individual's net initial premium to the risk reserve. This premium may be larger or smaller than the previous annuity payments, depending on the length of time the annuitant has survived. The risk reserve $R(t)$, with $R(0) = z$, is thus

$$R(t) = z - (p_1 - \eta_1)t + Y(t) \equiv z - \pi_2 t + Y(t), \qquad (4.36)$$

where $Y(t)$ is the aggregate of the annuity values paid to the risk reserve on account of deaths up to time t, the distribution function of individual annuity premiums is $B(y)$, $0 \leq y \leq \infty$, corresponding to $1 - P(-y)$ of the positive claim situation, and $p_1 = \int_0^\infty y \, dB(y)$.

Corresponding to relation (2.33), the aggregate addition to $R(0)$ from annuity premiums paid to the risk-reserve has the d.f.

$$F(y, t) = \sum_{j=0}^{\infty} e^{-t} \frac{t^j}{j!} B^{j*}(y) \qquad (4.37)$$

and an expression for $u(x, t)$ can be derived without requiring the prior calculation of $u(0, t)$.

Consider the risk process commencing at time zero with an initial risk reserve of z. The "curve" of $R(t)$ is a straight line sloping downward (with gradient π_2) from the point $(0, z)$ toward the horizontal t-axis until the first death occurs. It then moves vertically upward by an amount y_1, the single premium paid to the risk reserve at that time. The curve then continues its downward course with slope π_2. We require the probability that the curve has not crossed the t-axis by a given time τ.

This probability has a direct analog in queueing theory, where $\pi_2 = 1$. The ordinate $R(t)$ there represents the latent service demand at time t, namely, the amount of time that will be used by the server in attending to customers already in the queue (or being attended to) at time t. Any new customer joining the queue at time τ increases the aggregate service demand by an amount y determined by the d.f. $B(y)$. At such a point τ the "curve" of $R(t)$ jumps upward by the amount y. Between new-customer time-points $R(t)$ is a straight line downward at an angle of $45°$ to the t-axis. If the remaining service time required for the customer receiving attention at time $t = 0$ is z, the probability that $R(t)$ will not become zero by time t thereafter is $u(z, t)$, and the interval t until the curve $R(t)$ first crosses the time axis is called the busy period.

Suppose n deaths have occurred up to time $\tau \leq t$ and that the curve $R(\cdot)$ crosses the time axis between τ and $\tau + d\tau$. This means that

$$R(\tau) = R(0) - \pi_2\tau + Y(\tau) = 0$$

or

$$Y(\tau) = \pi_2\tau - R(0) = \pi_2\tau - z,$$

and the probability that n annuity premiums paid to the risk-reserve will amount to $Y(\tau)$ is $dB^{n*}(\pi_2\tau - z)$. By an argument similar to that used in deriving (4.8) the proportion of processes in which the curve of $R(\cdot)$ has reached the time axis for the first time at τ is $z/\pi_2\tau$. Hence the probability of n deaths and "first" ruin at time τ is

$$e^{-\tau} \frac{\tau^n}{n!} \frac{z}{\pi_2\tau} d_\tau B^{n*}(\pi_2\tau - z),$$

and, since this "first" ruin could have occurred at any τ-value between z/π_2 and t, the required probability of ruin sometime before time t is

$$v(z, t) = 1 - u(z, t) = \sum_{n=0}^{\infty} \int_{\tau=z/\pi_2}^{t} e^{-\tau} \frac{\tau^n}{n!} \frac{z}{\pi_2\tau} d_\tau B^{n*}(\pi_2\tau - z)$$

$$= \frac{z}{\pi_2} \sum_{n=0}^{\infty} \int_0^{\pi_2 t - z} e^{-(u+z)/\pi_2} \frac{(u + z/\pi_2)^{n-1}}{n!} d B^{n*}(u). \quad (4.38)$$

This result was derived for the queueing situation by Prabhu (1960) and later (1961) applied by him to the ruin of an annuity company.

Notice how similar this formula is to (4.8) but that it corresponds to the far more complicated formula (4.16) of the positive risk-sums case.

The L.T. of $v(z, t)$ is interesting. For convenience we write $\lambda = \Pi_2^{-1}$; then

$$\int_{\lambda z}^{\infty} e^{-st} v(z, t)\, dt$$

$$= \sum_{n=0}^{\infty} \frac{\lambda z}{n!} \int_{\lambda z}^{\infty} e^{-st}\, dt \int_{\lambda z}^{t} e^{-\tau} \tau^{n-1}\, d_\tau B^{n*}(\lambda^{-1}\tau - z)$$

$$= \sum_{n=0}^{\infty} \frac{\lambda z}{n!} \int_{\lambda z}^{\infty} e^{-\tau} \tau^{n-1}\, d_\tau B^{n*}(\lambda^{-1}\tau - z) \int_{\tau}^{\infty} e^{-st}\, dt$$

$$= \sum_{n=0}^{\infty} \frac{\lambda^n z}{n! s} \int_{z}^{\infty} e^{-\lambda(s+1)x} x^{n-1}\, d_x B^{n*}(x - z)$$

$$= s^{-1} e^{-\lambda(s+1)z} + \sum_{n=1}^{\infty} \frac{\lambda^n}{n!} \left[-\frac{d}{d(\lambda s + 1)} \right]^{n-1} \frac{z}{s} e^{-\lambda(s+1)z} \int_{0}^{\infty} e^{-\lambda(s+1)y}\, dB^{n*}(y)$$

$$= s^{-1} e^{-\lambda(s+1)z} + \sum_{n=1}^{\infty} \frac{\lambda^n}{n!} \left[-\frac{d}{d(\lambda s + 1)} \right]^{n-1} \frac{z}{s} e^{-\lambda(s+1)z} [\beta(\overline{\lambda s + 1})]^n, \qquad (4.39)$$

where

$$\beta(s) = \int_{0}^{\infty} e^{-sy}\, dB(y).$$

Now Lagrange's theorem (Whittaker and Watson, 1927) states that if the inequality $|t\, \phi(x)| < |(x - a)|$ is satisfied for all complex x on a contour C, on and within which $\phi(\cdot)$ is analytic, then the function $f(\cdot)$ which is analytic on and within C can be expanded by the formula

$$f(\zeta) = f(a) + \sum_{n=1}^{\infty} \frac{t^n}{n!} \frac{d^{n-1}}{da^{n-1}} \{f'(a)[\phi(a)]^n\}, \qquad (4.40)$$

where $\zeta = a + t\, \phi(\zeta)$ and this equation is satisfied by a single value of ζ in the interior of C.

Formally substituting in this theorem $a = -\lambda(s + 1), f(a) = s^{-1} e^{-\lambda(s+1)z}$, $t = \lambda$, and $\phi(\cdot) = \beta(\cdot)$, we obtain

$$\int_{\lambda z}^{\infty} e^{-st} v(z, t)\, dt = s^{-1} e^{-z\zeta} \quad \text{with} \quad \zeta = (s + 1)\pi_2^{-1} - \beta(\zeta). \qquad (4.41)$$

This relation (in L.S.T. form) was first given by Segerdahl (1939), who used it to find the moments of the distribution of the time of ruin. Saxén (1948) continued this work on $v(z, t)$ for an annuity fund and Arfwedson (1955) provided an asymptotic formula for $v(z, t)$ and explicit results for the cases in which (a) the annuity reserves are all equal and (b) $B(y) = 1 - e^{-y}$. The foregoing relation first appeared in queueing theory in Beneš (1957).

A difficult ruin problem that appears to have no analog in queueing theory is the general case in which both upward and downward jumps can occur in $R(t)$. An expression for the L.T. of $v(x, t)$ in this case is given by Beekman (1966)* and earlier work and references are contained in Arfwedson (1954–1955) and Cramér (1955). No general numerical results appear to have been published.

Illustration. If the distribution of annuity premiums received by the risk reserve is (negative) exponential with mean unity,

$$B(y) = 1 - e^{-y}, \qquad 0 \leq y < \infty,$$

and from (2.34)

$$B^{n*}(y) = 1 - e^{-y} \sum_{j=0}^{n-1} \frac{y^j}{j!},$$

so that

$$dB^{n*}(y) = e^{-y} \frac{y^{n-1}}{(n-1)!} \, dy, \qquad n = 1, 2, \ldots,$$

and

$$dB^{0*}(y) = \begin{cases} 0, & y < 0, \\ 1, & y \geq 0. \end{cases}$$

Hence

$$v(z, t) = e^{-z/\pi_2} + \sum_{n=1}^{\infty} z\pi_2^{-1} \int_{\tau=z}^{\pi_2/t} e^{-\tau/\pi_2} \frac{(\tau/\pi_2)^{n-1}}{n!} e^{-(\tau-z)} \frac{(\tau - z)^{n-1}}{(n-1)!} \, d(\tau - z)$$

$$= e^{-z/\pi_2} + z\pi_2^{-1} e^{-z/\pi_2} \int_0^{\pi_2/t-z} e^{-(1+\pi_2^{-1})\tau} I_0'\left(2\sqrt{\tau \frac{\tau + z}{\pi_2}}\right) d\tau. \qquad (4.42)$$

Alternatively the L.T. of $v(z, t)$ can be obtained from Segerdahl's relation (4.41).

THE PROBABILITY OF ULTIMATE RUIN

We have already mentioned that the right-hand side of (4.15) vanishes when $t = \infty$. We then have

$$u(x, \infty) = u(0, \infty) + \int_0^x u(x - y, \infty) h(y) \, dy, \quad \text{with} \quad h(y) = \frac{1 - P(y)}{\pi_1}$$

or, in the notation we have already used,

$$\phi(x) = \phi(0) + \int_0^x \phi(x - y) h(y) \, dy$$

and, using (4.11) with $p_1 = 1$ so that $\eta_1 = \eta$,

$$\phi(x) = \frac{\eta}{\pi_1} + \int_0^x \phi(x - y) h(y) \, dy. \qquad (4.43)$$

* See also Takács (1967).

This equation was first derived by Cramér (1926) in the form (4.61) and is an example of the "renewal equation" which we consider below. Let us first consider how to solve it by using Laplace transforms.

As before, write

$$\Phi(s) = \int_0^\infty e^{-sx}\,\phi(x)\,dx$$

and take the L.T. of (4.48). The result is

$$\Phi(s) = (s\pi_1)^{-1}\eta + \int_0^\infty e^{-sx}\,dx \int_0^x \phi(x - y)\,h(y)\,dy$$

$$= (s\pi_1)^{-1}\eta + \int_0^\infty e^{-sy}\,h(y)\,dy \int_y^\infty e^{-s(x-y)}\,\phi(x - y)\,dx$$

$$= (s\pi_1)^{-1}\eta + \int_0^\infty e^{-sy}\,h(y)\,dy \int_0^\infty e^{-sz}\,\phi(z)\,dz$$

$$= (s\pi_1)^{-1}\eta + \chi(s)\,\Phi(s),$$

so that from (4.26)

$$\Phi(s) = \frac{\eta}{\pi_1 s - 1 + \pi(s)}\,. \tag{4.44}$$

This relation was derived explicitly by Khintchine (1932) as the characteristic function of the waiting time in queueing theory, but essentially the same result is found in Lundberg (1928, 1930), Cramér (1930), and Pollaczek (1930).

Illustration.   Let
$$P(y) = 1 - e^{-y}, \qquad 0 < y < \infty;$$
then $p_1 = 1$ and
$$\pi(s) = \int_0^\infty e^{-(s+1)y}\,dy = \frac{1}{s + 1}$$
and (4.44) gives

$$\Phi(s) = \frac{\eta}{(1 + \eta)s - 1 + 1/(s + 1)} = \frac{\eta(s + 1)}{s[(1 + \eta)s + \eta]}$$

$$= \frac{1}{s} + \frac{\eta\left(1 - \dfrac{\eta}{1 + \eta}\right)\Big/\left(-\dfrac{\eta}{1 + \eta}\right)}{(1 + \eta)s + \eta} \qquad \text{by "partial fractions"}$$

$$= \frac{1}{s} - \frac{1}{(1 + \eta)s + \eta} = \frac{1}{s} - \frac{1}{(1 + \eta)[s + \eta/(1 + \eta)]}$$

$$= \int_0^\infty e^{-sx}\,dx - \frac{1}{1 + \eta}\int_0^\infty \exp\left[-\left(s + \frac{\eta}{1 + \eta}\right)x\right]dx.$$

On inverting this L.T.,

$$\phi(x) = 1 - \frac{1}{1 + \eta} e^{-\eta x/(1+\eta)}$$

In queueing theory the more general (renewal process) form of (4.43), applicable when $f(t)$, the density of independent interarrival times, corresponding to our density of interclaim times and given by $p_0''(t)/\lambda$ (McFadden, 1962; see Appendix A) is no longer $\lambda e^{-\lambda t}$, has been solved for some special cases. In the risk situation this general form, obtainable from (4.17) by replacing $e^{-\tau}$ with $p_0''(t)/\lambda$ and letting $t \to \infty$, may be written

$$\phi(x) = \int_0^\infty \phi(x + z)\, dz \int_0^\infty w(y + z)\, dP(y) + \int_0^x \phi(x - z)\, dz \int_0^\infty w(\tau)\, d_\tau P(z + \tau),$$

where

$$w(\tau) = \frac{1}{\lambda \pi_1} p_0'' \frac{\tau}{(\pi)}\, .$$

This equation has an explicit solution when $w(\cdot)$ and $P'(\cdot)$ are both gamma densities with integer indices n and m and means n/κ and m/μ, respectively, such that $m/\mu - n/\kappa < 0$ if $\phi(x)$ is to be nonzero. By giving one or both of these indices the special values unity and infinity, respectively, nine different relations for $\phi(\cdot)$ result. The reader is referred to Syski (1960, pp. 307–326) for details; there $\phi(x)$ is written $W(t)$ and is the probability that a customer will have a wait for service less than t if he joins the queue after it has existed for an infinite period (the "equilibrium" or "steady state" case to distinguish it from the "transient" state soon after the first customer arrives). More general theorems on the behavior of this (Lindley) equation are quoted by Syski (*loc. cit.*, pp. 326–328).

Our success with the inversion of the particular case of $\Phi(s)$ for equal sums at risk encourages us to attempt the same procedure again.

$$\Phi(s) = \eta(\pi_1 s - 1)^{-1} \left[1 + \frac{\pi(s)}{\pi_1 s - 1} \right]^{-1}$$

$$= \eta \sum_{j=0}^\infty (-1)^j (\pi_1 s - 1)^{-j-1} [\pi(s)]^j$$

$$= \eta \sum_{j=0}^\infty \frac{(-1)^j}{j!} (\pi_1 s - 1)^{-j-1} [\pi(s)]^j \int_0^\infty x^j e^{-x}\, dx$$

$$= \eta \sum_{j=0}^\infty \frac{(-1)^j}{j!} [\pi(s)]^j \int_0^\infty v^j e^{-v(\pi_1 s - 1)}\, dv$$

$$= \eta \sum_{j=0}^\infty \frac{(-1)^j}{j!} \pi_1^{-j-1} [\pi(s)]^j \int_0^\infty u^j e^{u/\pi_1 - su}\, du.$$

Now $[\pi(s)]^j$ is the L.S.T. of $P^{j*}(y)$ and the integral last written is the L.T. of $u^j e^{u/\pi}$. Since the product of two L.T.'s is the L.T. of the convolution of the corresponding functions

$$\phi(x) = \eta \sum_{j=0}^{\infty} \frac{(-1)^j}{j!} \pi_1^{-j-1} \int_0^x (x - y)^j e^{(x-y)/\pi_1} \, dP^{j*}(y). \tag{4.45}$$

An alternative procedure is instructive. We have

$$\Phi(s) = \frac{\eta/\pi_1 s}{1 - [1 - \pi(s)]/\pi_1 s} = \eta \sum_{j=0}^{\infty} \pi_1^{-j-1} s^{-1} \left[\frac{1 - \pi(s)}{s} \right]^j.$$

Now $1/s - \pi(s)/s$ is the L.T. of $1 - P(y) = Q(y)$ and $[1 - \pi(s)/s]^j$ is thus the L.T. of $Q^{j*}(y)$ defined by

$$Q^{j*}(y) = \int_0^y Q^{(j-1)*}(y - x) \, Q(x) \, dx, \qquad j = 2, 3, 4, \ldots .$$

Finally $s^{-1}\{[1 - \pi(s)]/s\}^j$ is the L. T. of $\int_0^y Q^{j*}(y - x) \, dx$, since s^{-1} is the L.T. of unity, so that

$$\phi(x) = \eta \sum_{j=0}^{\infty} \pi_1^{-j-1} \int_0^x Q^{j*}(y - x) \, dx, \tag{4.46}$$

a formula derived by Beekman as recently as 1966. Another alternate form for $\phi(x)$ is obtained by taking the limit of (4.16) as $t \to \infty$ and utilizing (2.33), namely,

$$u(z, \infty) = \phi(x) = 1 - \eta \int_0^{\infty} \sum_{j=0}^{\infty} e^{-\tau} \frac{\tau^j}{j!} \, d_\tau P^{j*}(z + \pi_1 \tau). \tag{4.47}$$

Note that it follows from (4.45) or (4.46) that $\phi(x) = 0$ (ultimate ruin is certain) when $\eta = 0$.

THE EFFECTS OF IMPOSING A CEILING ON THE RISK RESERVE OR OF USING A VARIABLE LOADING

The original reason for calculating $\phi(x)$ rather than $u(x, t)$ was mathematical and numerical convenience. The concept of an indefinite future for a risk business coupled with the resulting infinite risk reserve [if $\eta \neq 0$] was, however, criticized as unrealistic. At first sight the continued growth of the risk reserve could be prevented by setting an upper limit on it. Suppose that whenever the risk reserve attains an amount Z, further premiums are used only to keep it "topped up" to that amount. It can be shown that such a procedure implies the eventual ruin of the risk business.

Suppose that there is a claim size $Y = y_0$ such that $1 - P(y_0) = \alpha \neq 0$. Choose n such that $Z + \pi_1 t - n y_0 < 0$. Now the probability that exactly n claims will occur in a finite period t is $p_n(t) \neq 0$ and the probability that every one of them will exceed y_0 is α^n. In an infinite period there is an infinite number of such finite periods of length t and ruin will occur in a proportion $p_n(t)\alpha^n$ of them. Ruin is thus certain in an infinite period of time.

Instead of attempting to curtail the risk reserve directly, we may consider the possibility of adjusting the risk-loading upward or downward, depending on the size of the risk reserve (Lundberg, 1926 and earlier). If we differentiate (4.48) with respect to x, remembering that $h(x) = \pi_1^{-1}[1 - P(x)]$,

$$\phi'(x) = \phi(0)\, h(x) + \int_0^x \phi'(x - y)\, h(y)\, dy$$

$$= -\pi_1^{-1}\, \phi(x) - \pi_1^{-1} \int_0^x \phi(x - y)\, dP(y),$$

so that, on changing π_1 to $\pi_1(x) = p_1 + \eta(x)$, where $\eta'(x) < 0$,

$$-\pi_1(x)\, \phi'(x) + \phi(x) = \int_0^x \phi(x - y)\, dP(y).$$

Even simple assumptions like $\eta(x) = \eta e^{-\beta x}$ do not produce explicit relations for $\phi(\cdot)$ or $\Phi(\cdot)$, but Davidson (1946) has proved that if $\eta(x) = 0$ for $x \geq a$ ruin is certain no matter how large $\eta(x)$ is for $x < a$. He has also considered examples in which $\eta(x)$ is a positive constant for $x \geq a$.

A special case of the variable loading is one in which interest is credited on the risk reserve with force δ. We then have $\pi_1(x) = \pi_1 + \delta x$ and an explicit result for $\phi(x)$ is available in the case $P(y) = 1 - e^{-y}$ (Segerdahl, 1942).

THE DIRECT SOLUTION OF THE RENEWAL EQUATION

Equation 4.43 is a special case of the "renewal equation" for $Z(x)$, namely,

$$Z(x) = z(x) + \int_0^x Z(x - y)\, dF(y), \tag{4.48}$$

where both $Z(\cdot)$ and $z(\cdot)$ vanish on $(-\infty, 0)$, and $F(\cdot)$ is a distribution function with $F(0) = 0$. We note that the upper limit of the integral could be written as ∞ without changing the equation. The distribution function $F(x)$ would be a *probability* distribution function if $F(\infty) = 1$; in general we only assume that $F(\infty) \leq 1$.

In renewal theory $Z(x)$ is, for example, the number of female births at time x and $F'(y)$ is the expected number of new-born daughters who survive y years and then give birth to a female child. The resulting equation

$$Z(x) = \int_0^\infty Z(x - y)\, dF(y)$$

can be rewritten in the form (4.48) if it is supposed that values of $Z(x)$ for negative x are known. The considerable literature on "self-renewing aggregates" is indicated by Saxer (1958) in his textbook and an elegant mathematical review of the subject is found in Feller (1966). The stochastic (point) process that leads naturally to (4.48) is described in the Appendix A.

Write, $n = 1, 2, 3, \ldots$

$$F^{n*}(x) = \int_0^x F^{(n-1)*}(x - y)\, dF(y), \qquad F^{1*}(x) = F(x), \qquad (4.49)$$

$$F^{0*}(x) = \begin{cases} 0, & x \le 0, \\ 1, & x > 0. \end{cases}$$

The function $U(x) \equiv \sum_{n=0}^{\infty} F^{n*}(x)$ plays an important role in the solution of (4.48), and it is necessary to show that it is finite for every x. Since $F(x)$ is a d.f. $F^{n*}(x) \le [F(x)]^n$ and $U(x)$ converges as a geometric progression with common ratio less than unity for values of x for which $F(x) < 1$. However, whatever x is there exists a finite integer r such that $F^{r*}(x) < 1$. Furthermore, since $F^{r*}(x)$ is monotonic downward with increasing n

$$F^{kr*}(x) + F^{(kr+1)*}(x) + \cdots + F^{(\overline{k+1}\ r-1)*}(x) \le rF^{kr*}(x)$$

and

$$U(x) \equiv \sum_{n=0}^{\infty} F^{n*}(x) \le r\sum_{k=0}^{\infty} F^{rk*}(x) \le r\sum_{k=0}^{\infty} [F^{r*}(x)]^k = r[1 - F^{r*}(x)]^{-1}$$

and this is bounded for all x. We state that if $z(\cdot)$ is bounded the only bounded solution of (4.48) vanishing on $(-\infty, 0)$ is

$$Z(x) = \int_0^x z(x - y)\, dU(y). \qquad (4.50)$$

For (4.50) is equivalent to

$$Z(x) = \sum_{n=0}^{\infty} \int_0^x z(x - y)\, dF^{n*}(y)$$

$$= z(x) + \sum_{n=1}^{\infty} \int_{y=0}^x z(x - y)$$

$$\times \left[\int_{w=0}^y d_y F^{(n-1)*}(y - w)\, dF(w) + F^{(n-1)*}(0)\, dF(y) \right]$$

$$= z(x) + \int_{y=0}^x z(x - y) \int_{w=0}^y d_y U(y - w)\, dF(w) \quad \text{since} \quad F^{(n-1)*}(0) = 0$$

$$= z(x) + \int_{w=0}^x dF(w) \int_{y=w}^x z(x - y)\, d_y U(y - w)$$

$$= z(x) + \int_0^x dF(w) \int_0^{x-w} z(x - w - s)\, dU(s)$$

$$= z(x) + \int_0^x Z(x - w)\, dF(w), \qquad \text{by (4.50)}$$

This is (4.48). Hence (4.50) is a solution of (4.48).

If $W(x)$ is the difference between two such solutions, then from (4.48)

$$W(x) = \int_0^x W(x - y)\, dF(y),$$

and on repeated substitution of this solution into its right-hand member

$$W(x) = \int_0^x W(x - y)\, dF^{n*}(y), \qquad n = 1, 2, 3, \ldots .$$

As $n \to \infty$ $F^{n*}(x) \to 0$, all x, and thus $W(x) = 0$. The solution (4.50) is therefore unique.

We may now apply these general results to (4.43), namely,

$$\phi(x) = \frac{\eta_1}{\pi_1} + \int_0^x \phi(x - y)\, h(y)\, dy, \qquad x \geq 0,$$

by putting

$$\int_0^x h(y)\, dy = F(x),$$

thus implying that $F(0) = 0$ and $F(\infty) = p_1/\pi_1 < 1$, and writing η_1/π_1 for $z(x)$. The solution (4.50) then becomes

$$\phi(x) = \frac{\eta_1}{\pi_1} \int_0^x dU(y) = \frac{\eta_1}{\pi_1}\, U(x), \tag{4.51}$$

where

$$U(x) = \sum_{n=0}^{\infty} F^{n*}(x)$$

and

$$F^{n*}(x) = \int_0^x F^{(n-1)*}(x - y)\, h(y)\, dy. \tag{4.52}$$

Illustration. Suppose that

$$P(y) = 1 - e^{-y}, \qquad 0 < y < \infty,$$

so that

$$p_1 = \int_0^{\infty} y\, dP(y) = \int_0^{\infty} y e^{-y}\, dy = 1,$$

$$h(y) = \frac{1 - P(y)}{p_1 + \eta_1} = \frac{e^{-y}}{1 + \eta}, \qquad z(x) = \frac{\eta}{1 + \eta},$$

and

$$F^{1*}(x) = F(x) = \int_0^x h(y)\, dy = \frac{1}{1 + \eta} \int_0^x e^{-y}\, dy = \frac{1 - e^{-x}}{1 + \eta}.$$

Then from (4.52)

$$F^{2*}(x) = \int_0^x \frac{1 - e^{-x+y}}{1 + \eta} \frac{e^{-y}}{1 + \eta}\, dy = \left(\frac{1}{1 + \eta}\right)^2 (1 - e^{-x} - xe^{-x}),$$

$$F^{3*}(x) = (1 + \eta)^{-3} \int_0^x [1 - e^{-x+y} - (x - y)e^{-x+y}]e^{-y}\, dy$$

$$= (1 + \eta)^{-3}\left(1 - e^{-x} - xe^{-x} - \frac{x^2}{2!}e^{-x}\right),$$

and, generally, $n = 1, 2, 3, \ldots$,

$$F^{n*}(x) = (1 + \eta)^{-n}\left[1 - e^{-x}\sum_{j=0}^{n-1}\frac{x^j}{j!}\right] = (1 + \eta)^{-n}e^{-x}\sum_{j=n}^{\infty}\frac{x^j}{j!}.$$

Hence

$$U(x) = 1 + \sum_{n=1}^{\infty} F^{n*}(x) = 1 + \sum_{n=1}^{\infty}(1 + \eta)^{-n}e^{-x}\sum_{j=n}^{\infty}\frac{x^j}{j!}$$

$$= 1 + e^{-x}\sum_{j=1}^{\infty}\frac{x^j}{j!}\sum_{n=1}^{j}(1 + \eta)^{-n}$$

$$= 1 + e^{-x}\sum_{j=1}^{\infty}\frac{x^j}{j!}(1 + \eta)^{-1}\frac{1 - (1 + \eta)^{-j}}{1 - (1 + \eta)^{-1}}$$

$$= 1 + \frac{e^{-x}}{\eta}[(e^x - 1) - (e^{x/(1+\eta)} - 1)]$$

$$= 1 + \frac{1}{\eta}[1 - e^{-\eta x/(1+\eta)}], \qquad 0 < x < \infty,$$

and (4.51) becomes

$$\phi(x) = \frac{\eta}{1 + \eta}\left[\frac{1 + \eta}{\eta} - \frac{1}{\eta}e^{-\eta x/(1+\eta)}\right]$$

$$= 1 - \frac{1}{1 + \eta}e^{-\eta x/(1+\eta)}.$$

as already obtained when illustrating (4.44).

If, for example, the initial risk reserve amounted to 100 times the expected unit claim, then, with a risk loading of 10% on the unit pure premium paid in unit (operational) time,

$$\psi(100) = \left(\frac{1}{1.1}\right)\exp\left(\frac{-10}{1.1}\right) = 10^{-4} \times \frac{1.126855}{1.1} = 0.000102.$$

ASYMPTOTIC VALUE OF THE PROBABILITY OF ULTIMATE RUIN

In the application of the general renewal equation (4.53) to our special case we have $F(\infty) = p_1/\pi_1 < 1$. In order, therefore, to approximate to

$\phi(x)$ with x large, we reconsider the renewal equation with

$$F(\infty) = k < 1.$$

Consider the behavior of $Z(x)$ as $x \to \infty$. The relation

$$F^{n*}(x) = \int_0^x F^{(n-1)*}(x - y) \, dF(y)$$

indicates that as x increases the contribution of the integrand comes from the "constant" $F^{(n-1)*}(\infty)$ and the total mass of $F(\cdot)$, namely k. Hence

$$F^{n*}(x) \to k \, F^{(n-1)*}(x) \to k^2 \, F^{(n-2)*}(x) \to \cdots \to k^n$$

and thus as $x \to \infty$

$$U(x) = \sum_{n=0}^{\infty} F^{n*}(x) \to \sum_{n=0}^{\infty} k^n = (1 - k)^{-1}.$$

Relation 4.50 then provides

$$Z(x) = \int_0^x z(x - y) \, dU(y) \to z(\infty)(1 - k)^{-1}. \tag{4.53}$$

When $Z(x) = \phi(x)$, $F(x) = \int_0^x h(y) \, dy$ and $z(x) = \phi(0)$,

the equation preceding (4.43) has a solution that tends to

$$\phi(\infty) = \phi(0)(1 - p_1/\pi_1)^{-1} = 1,$$

so that $\phi(0) = \eta/\pi_1$. This was the "classic" method of finding $\phi(0)$ before Takács' (1967) theorem became available. It is of little use, however, in the derivation of an asymptotic relation for $\phi(x)$.

Let us now suppose that $F(\infty) = 1$ and derive an asymptotic expression for $Z(x)$. We proceed in two stages: in the first we suppose $z(\cdot)$ is a true d.f. and obtain an asymptotic value for $Z(x)$; then we rewrite (4.48) so that $z(x)$ becomes a d.f. and modify the preceding asymptotic result. Having thus derived an asymptotic expression for $Z(x)$ in (4.48) with $F(\infty) = 1$, we finally redefine $Z(\cdot)$, $z(\cdot)$, and $F(\cdot)$ in such a way that the modified $F(\cdot)$ has an upper bound of less than unity. The asymptotic result for this case then applies directly to $\phi(x)$.

Consider first the special case of the general renewal equation (4.48) in which

$$Z(t) = F(t) + \int_0^t Z(t - y) \, dF(y) \tag{4.54}$$

and $F(\cdot)$ is a proper distribution function on $[0, \infty)$.

Suppose that $$S_j = X_1 + X_2 + \cdots + X_j$$

is the sum of j identical independent random variables each distributed as $F(\cdot)$ and put $\mathcal{E}(X) = \mu$, a finite number. Write $N(t)$, a random variable, for the number of values of S_j that do not exceed t ($j = 1, 2, \ldots$). Since all $X_j \geq 0$, the number $N(t)$ is the largest value of n such that $S_n \leq t$ and $S_{n+1} > t$.

By the strong law of large numbers with probability one there exists a number n_1 such that with $\varepsilon > 0$ arbitrarily small

$$\mu - \varepsilon < \frac{S_{n_1}}{n_1} < \mu + \varepsilon;$$

but t can be chosen so large that $N(t) = n_2 > n_1$, so that

$$S_{n_2} \leq t \quad \text{but} \quad S_{n_2+1} > t.$$

Hence

$$\mu - \varepsilon < \frac{S_{n_2}}{n_2} \leq \frac{t}{n_2} \quad \text{and} \quad \mu + \varepsilon > \frac{S_{n_2+1}}{n_2 + 1} > \frac{t}{n_2 + 1},$$

leading to

$$\frac{1}{\mu + \varepsilon} - \frac{1}{t} < \frac{n_2}{t} < \frac{1}{\mu - \varepsilon}$$

or

$$\lim_{t \to \infty} \frac{N(t)}{t} = \frac{1}{\mu}, \quad \text{with probability one.}$$

Now

$$\mathcal{E}\{N(t)\} = \sum_{j=1}^{\infty} j \, \mathcal{F}\{S_j \leq t \quad \text{and} \quad S_{j+1} > t\}$$

$$= \sum_{j=0}^{\infty} j \, \mathcal{F}(S_j \leq t) \cdot \mathcal{F}(S_{j+1} > t \mid S_j \leq t)$$

$$= \sum_{j=0}^{\infty} j \, \mathcal{F}(S_j \leq t)\{1 - \mathcal{F}(S_{j+1} \leq t \mid S_j \leq t)\}$$

$$= \sum_{j=0}^{\infty} j\{\mathcal{F}(S_j \leq t) - \mathcal{F}(S_{j+1} \leq t)\}$$

$$= \sum_{j=0}^{\infty} \{j \, \mathcal{F}(S_j \leq t) - (j + 1) \, \mathcal{F}(S_{j+1} \leq t) + \mathcal{F}(S_{j+1} \leq t)\}$$

$$= \sum_{j=1}^{\infty} \mathcal{F}(S_j \leq t)$$

$$= \sum_{n=1}^{\infty} F^{n*}(t) = \sum_{n=0}^{\infty} \int_0^t F(t - y) \, dF^{n*}(y),$$

and this is $Z(t)$, the solution of the renewal equation (4.54).

Hence, on taking expected values of both sides of the limiting relation for $N(t)/t$,

$$\frac{1}{\mu} = \lim_{t \to \infty} \frac{N(t)}{t} = \lim_{t \to \infty} \frac{\mathcal{E}\{N(t)\}}{t} = \lim_{t \to \infty} \frac{Z(t)}{t}.$$

We may thus write

$$Z(t) \sim \frac{t}{\mu} = \left[\int_0^\infty y \, dF(y) \right]^{-1} t. \tag{4.55}$$

Extension of this result to the more general equation (4.48) is effected as follows. First we integrate (4.48) between zero and x and divide the result by $\int_0^\infty z(t) \, dt$, which we assume is finite. Writing

$$Z_0(x) = \frac{\displaystyle\int_0^x Z(t) \, dt}{\displaystyle\int_0^\infty z(t) \, dt},$$

we obtain the result

$$Z_0(x) = F_0(x) + \frac{\displaystyle\int_0^x dt \int_0^t Z(t - y) \, dF(y)}{\displaystyle\int_0^\infty z(t) \, dt}$$

$$= F_0(x) + \int_0^x Z_0(x - y) \, dF(y), \tag{4.56}$$

where $F_0(x) = \int_0^x z(t) \, dt / \int_0^\infty z(t) \, dt$ and is thus a proper d.f. on $(0, \infty)$. Then we stipulate that X_0, a new first variable in S_j, has a distribution function $F_0(\cdot)$ that differs from that of $X_1, X_2, \ldots$ (and has a finite mean). The strong law of large numbers still holds and the inequalities of our proof remain valid; furthermore the expression for $\mathcal{E}\{N(t)\}$ is modified without difficulty.

Our general result for the case in which $F(\infty) = 1$ thus becomes

$$Z_0(t) \sim \frac{t}{\mu} \tag{4.57}$$

and

$$Z_0'(t) = \frac{Z(t)}{\displaystyle\int_0^\infty z(t) \, dt} \sim \frac{1}{\mu},$$

so that

$$Z(t) \sim \frac{\displaystyle\int_0^\infty z(t) \, dt}{\displaystyle\int_0^\infty y \, dF(y)}. \tag{4.58}$$

This agrees with (4.55) when $z(x) = F(x)$.

We may now apply this result to the case in which the distribution function $F(\cdot)$ is defective by introducing a number κ such that

$$\int_{-\infty}^\infty e^{\kappa x} \, dF(x) = 1, \qquad \kappa > 0,$$

and then

$$F^+(x) = \int_0^x e^{\kappa y}\, dF(y) \tag{4.59}$$

is a probability distribution function with $F^+(\infty) = 1$.

Multiplying (4.48) by $e^{\kappa x}$, we obtain

$$e^{\kappa x} Z(x) = e^{\kappa x} z(x) + \int_0^x e^{\kappa(x-y)} Z(x-y) e^{\kappa y}\, dF(y)$$

or, say,

$$Z^+(x) = z^+(x) + \int_0^x Z^+(x-y)\, dF^+(y)$$

and the asymptotic result (4.58) provides

$$Z^+(x) \to \frac{\displaystyle\int_0^\infty z^+(y)\, dy}{\displaystyle\int_0^\infty y\, dF^+(y)} = \frac{\displaystyle\int_0^\infty e^{\kappa y} z(y)\, dy}{\displaystyle\int_0^\infty y e^{\kappa y}\, dF(y)}. \tag{4.60}$$

This applies generally to (4.48) in the case in which $F(\infty) < 1$.

To apply these results to (4.43) we first write it as

$$1 - \phi(x) = \psi(x) = \int_0^\infty h(y)\, dy - \int_0^x \phi(x-y)\, h(y)\, dy$$

$$= \int_x^\infty h(y)\, dy + \int_0^x \psi(x-y)\, h(y)\, dy \tag{4.61}$$

and put $F(x) = \int_0^x h(y)\, dy$, $z(x) = \int_0^x h(y)\, dy$ and $Z(x) = \psi(x)$. Then κ is given by

$$\int_0^\infty e^{\kappa x} h(x)\, dx = 1, \tag{4.62}$$

and (4.60) yields

$$\psi(x) \to e^{-\kappa x} \frac{\displaystyle\int_0^\infty e^{\kappa y}\, dy \int_y^\infty h(z)\, dz}{\displaystyle\int_0^\infty y e^{\kappa y} h(y)\, dy} = e^{-\kappa x} \frac{\displaystyle\int_0^\infty h(z)\, dz \int_0^z e^{\kappa y}\, dy}{\displaystyle\int_0^\infty y e^{\kappa y} h(y)\, dy}$$

$$= e^{-\kappa x} \frac{\displaystyle\int_0^\infty (e^{\kappa z} - 1)\, h(z)\, dz}{\kappa \displaystyle\int_0^\infty y e^{\kappa y} h(y)\, dy}$$

$$= e^{-\kappa x} \frac{\displaystyle\int_0^\infty dP(y) \int_0^y (e^{\kappa z} - 1)\, dz}{\kappa \displaystyle\int_0^\infty dP(z) \int_0^z y e^{\kappa y}\, dy}$$

$$= \frac{p_0^+ - 1 - \kappa p_1}{\kappa p_1^+ - p_0^+ + 1}\, e^{-\kappa x}, \tag{4.63}$$

where

$$p_j^+ = \int_0^\infty y^j e^{\kappa y} \, dP(y), \qquad j = 0, 1.$$

By writing (4.62) in the form

$$\frac{1}{\kappa}(p_0^+ - 1) = p_1 + \eta_1$$

(4.63) is simplified to

$$\psi(x) \sim \frac{\eta_1}{p_1^+ - p_1 - \eta_1} \, e^{-\kappa x}, \tag{4.64}$$

a formula due to Lundberg (1926). The same author had obtained the inequality $\psi(x) < e^{-\kappa x}$ in an early paper (1909). The special case of (4.64) for sums at risk all equal to h was derived by Erlang (1909).

The relation (4.64) is equivalent to that preceding (4.34), which was derived by using Laplace transform theory. That method may be applied directly to the L.T. of $\phi(x)$ given by (4.44), the denominator of which has simple zeros at $s = 0$ and $s = s_1 > 0$. We then obtain

$$\phi(x) \sim 1 - \frac{\eta_1 e^{s_1 x}}{p_1' - \pi_1} \,,$$

where $s_1 = -\kappa$ and $p_1' = \int_0^\infty e^{-s_1 y} y \, dP(y)$. Operating in the same way on $[\Phi(s)]^k$, whose poles are of the kth order ($k = 1, 2, 3, \ldots$), we get

$$\phi^{k*}(x) \sim 1 + \frac{(-\eta_1)^k e^{s_1 x}}{(p_1' - \pi_1)^k} \,. \tag{4.65}$$

Illustrations.   As an illustration of (4.64), let us assume once again that

$$P(y) = 1 - e^{-y}, \qquad 0 < y < \infty.$$

Then

$$p_1 = 1 \quad \text{and} \quad h(y) = \frac{e^{-y}}{1 + \eta} \,,$$

so that (4.62) becomes

$$\int_0^\infty e^{\kappa x} e^{-x} \, dx = 1 + \eta;$$

that is,

$$\kappa = 1 - \frac{1}{1 + \eta} = \frac{\eta}{1 + \eta} \,, \qquad p_0^+ = \int_0^\infty e^{\kappa y} \, dP(y) = 1 + \eta,$$

and

$$p_1^+ = \int_0^\infty y e^{-y(1-\kappa)} \, dy = (1 - \kappa)^{-2} = (1 + \eta)^2.$$

Hence

$$\psi(x) \to \frac{1}{1 + \eta} \, e^{-\eta x/(1+\eta)},$$

which, as we have seen, is an exact (not only an asymptotic) result. Segerdahl (1957) has illustrated how κ and the factor multiplying $e^{-\kappa x}$ in (4.64) change with different assumptions about $P(y)$ and η. Writing

(i) $P(y) = \begin{cases} 0, & y < 1, \\ 1, & y \geq 1, \end{cases}$

(ii) $P(y) = 1 - e^{-y}, \quad 0 < y < \infty,$

(iii) $P'(y) = \begin{cases} 4.897954 \, e^{-5.514588y} + 4.503(y + 6)^{-2.75}, & 0 < y < 500, \\ 0, & y > 500, \end{cases}$

he obtained the results shown in Table 4.4 for three chosen κ-values.

Table 4.4

$10^4\kappa$	(i) $10^5\eta$	Factor	(ii) $10^5\eta$	Factor	(iii) $10^5\eta$	Factor
25	125	0.99916	251	0.9975	6380	0.8426
50	250	0.99833	503	0.9950	15769	0.6760
70	351	0.99767	705	0.9930	27322	0.5473

Although for the long-tailed distribution the probability of ruin is rather smaller for given κ there is a substantial increase in the risk loading necessary to achieve this size of probability.

REFERENCES

Andersen, E. S. (1957). "On the collective theory of risk in the case of contagion between the claims." *Trans. XV Intern. Cong. Actu.*, New York, **2**, 219–227.

Arfwedson, G. (1950). "Some problems in collective theory of risk." *Skand. Aktu. Tidskr.*. **33**, 1–38.

———— (1954–1955). "Research in collective risk theory." *Skand. Aktu. Tidskr.*, **37**, 191–223; **38**, 37–100.

———— (1957). "Notes on collective risk theory." *Skand. Aktu. Tidskr.*, **40**, 46–59.

Beekman, J. A. (1966). "Research on the collective risk stochastic process." *Skand. Aktu. Tidskr.*, **49**, 65–77.

Beneš, V. E. (1957). "On queues with Poisson arrivals." *Ann. Math. Statist.*, **28**, 670–677.

———— (1960). "Combinatory methods and stochastic Kolmogorov equations in the theory of queues with one server." *Trans. Amer. Math. Soc.*, **94**, 282–294.

Bhat, U. N. (1968). "A study of the queueing systems M/G/1 and GI/M/1." *Lecture Notes in Operations Research and Mathematical Economics*, **2**. Springer, Berlin.

Burman, J. P. (1946). "Sequential sampling formulae for a binomial population." *J. Roy. Statist. Soc. B*, **8**, 98–103.

Cox, D. R., and P. A. W. Lewis. *The Statistical Analysis of Series of Events*. Methuen, London.

Cramér, H. (1926). Review of Lundberg (1926). *Skand. Aktu. Tidskr.*, **9**, 223–245.

———— (1930). "On the mathematical theory of risk." *Försäkringsaktiebolaget Skandia*, 1855–1930. Stockholm.

———— (1955). "Collective risk theory: A survey of the theory from the point of view of the theory of stochastic processes." *Försäkringsaktiebolaget Skandia*, 1855–1955. Esselta, Centraltryckeriet, Stockholm.

Davidson, A. (1946). "Om ruinproblemet i den kollektiva riskteorin under antagande av variabel säkerhetsbelastning." *Försäkringsmatematiska Studier tillägnade Filip Lundberg*. Almqvist and Wiksells, Stockholm.

Erlang, A. K. (1909). "Sandsynlighedsregning og telefonsamtaler." *Nyt. Tidskr. Mat. B*, **20**, 33–40. [Translation in E. Brockmeyer, H. L. Halstrøm, and A. Jensen, (1948) *The Life and Works of A. K. Erlang*. Copenhagen Telephone Co., Copenhagen.]

Feller, W. (1968). *An Introduction to Probability Theory and Its Applications*, Vol. II. Wiley, New York.

Fürst, D. (1956). "La rovina dei giocatori nel caso di riserva limitata." *G. Ist. Ital. Attu.*, **19**, 63–83.

Fry, T. C. (1928, 1965). *Probability and Its Engineering Uses*. Von Nostrand, Princeton, N.J.

Khintchine, A. (1932), "Mathematisches über die Erwartung vor einem öffentlichen Schalter." *Matem. Sbornik*, **39**, 73–84.

Klinger, H. (1965). "Über die Verteilung des Ruinzeitpunktes bei beschränkter Risikoreserve." *Bl. Deuts. Gesell. Versich. Math.*, **7**, 389–407.

Lundberg, F. (1909), "Über die Theorie der Rückversicherung." *Ber. VI Intern. Kong. Versich. Wissens.*, **1**, 877–948.

———— (1926–1928). *Försakringsteknisk Riskutjämning*. I *Teori;* II *Statistik*. Englund, Stockholm.

———— (1930). "Über die Wahrscheinlichkeitsfunktion einer Risikenmasse." *Skand. Aktu. Tidskr.*, **13**, 1–83.

McFadden, J. A. (1962). "On the lengths of intervals in a stationary point process." *J. Roy. Statist. Soc.*, *B*, **24**, 364–382.

———— (1965). "The mixed Poisson process." *Sankhyā*, *A*, **27**, 83–92.

Pollaczek, F. (1930). "Über eine Aufgabe der Wahrscheinlichkeitstheorie." *Math. Zeit.*, **32**, 64–100, 729–750.

———— (1957), "Problèmes stochastiques posés par le phénomène de formation d'une queue d'attente à un guichet et par des phénonènes apparantés." *Mem. Sci. Math.* **136**, 1–123.

Prabhu, N. U. (1960). "Some results for the queue with Poisson arrivals." *J. Roy. Statist. Soc.*, *B*, **22**, 104–107.

———— (1961). "On the ruin problem of collective risk theory." *Ann. Math. Statist.*, **32**, 757–764.

Saxén, T. (1948). "On the probability of ruin in the collective risk theory for insurance enterprises with only negative risk sums." *Skand. Aktu. Tidskr.*, **31**, 199–228.

Saxer, W. (1958). *Versicherungsmathematik*, II. Springer, Berlin.

Seal, H. L. (1966). "The random walk of a simple risk business." *Astin Bull.*, **4**, 19–28.

Segerdahl, C.-O. (1939). *On Homogeneous Random Processes and Collective Risk Theory*. Almqvist and Wiksells, Uppsala.

——— (1942). "Über einige risikotheoretische Fragestellungen." *Skand. Aktu. Tidskr.*, **25**, 43–83.

——— (1955). "When does ruin occur in the collective theory of risk?" *Skand. Aktu. Tidskr.*, **38**, 22–36.

——— (1957). "If a risk business goes bankrupt, when does it occur? A basis for fixing net retention." *Trans. XV Intern. Cong. Actu.*, New York, **2**, 280–285.

Syski, R. (1960). *Introduction to Congestion Theory in Telephone Systems.* Oliver and Boyd, Edinburgh.

Takács, L. (1955). "Investigation of waiting time problems by reduction to Markov processes." *Acta. Math. Acad. Sci. Hungar.*, **6**, 101–128.

——— (1967). *Combinatorial Methods in the Theory of Stochastic Processes.* Wiley, New York.

Thomson, W. T. (1960). *Laplace Transformation.* Prentice-Hall, Englewood Cliffs, N.J.

Weesakul, B. (1961). "The random walk between a reflecting and an absorbing barrier." *Ann. Math. Statist.*, **32**, 765–769.

Whittaker, E. T., and G. N. Watson (1927). *A Course of Modern Analysis.* Cambridge University Press, Cambridge.

Wold, H. (1949). "Sur les processus stationnaires ponctuels." *Le Calcul des Probabilités et Ses Applications* (Lyon), C.N.R.S. **13**, 75–86.

CHAPTER 5

Premium Loading and Reinsurance

THE RISK PREMIUM LOADING

The loading η_1 added to κ_1 the expected claims (whether on an individual contract or on the aggregate portfolio) serves the double purpose of providing a profit and preventing certain ruin. For the time being we shall ignore the profit motive by assuming, for example, that all interest and investment gains on the risk reserve are paid to the company's stockholders. We may thus consider the purpose of η_1 to control the probability of ruin at a given level.*

An immediate method of calculating η_1 would be to fix $u(x, t)$, where x is the current (or initial) risk reserve and t is a suitably chosen period, at 0.99 or 0.999 (namely, near certainty) and to solve for η in (4.15) or (4.16). This could be done only by iterative methods and we know of no numerical example in the literature.

However, if t is allowed to increase without bound, it is possible to utilize the asymptotic formula (4.64).

Illustration. In the case of an annuity fund it is easy to see that (4.62) becomes

$$\int_0^\infty e^{-\kappa x} \frac{1 - B(x)}{\pi_2} \, dx = 1,$$

where $B(x)$ is $1 - P(-x)$, $0 < x < \infty$, and $\pi_2 = p_1 - \eta$. This relation is equivalent to

$$\int_0^\infty e^{-\kappa x} \, dB(x) = 1 - \kappa(p_1 - \eta). \tag{5.1}$$

It can also be shown that the factor multiplying the exponential in (4.64) is exactly (not only approximately) unity.

---

* We ignore the deliberate bias that may be introduced into κ_1, our estimate of κ_1.

Now O. Lundberg (1954) found that the variance of the sums at risk in a company making continuous payments of disability annuities was 2.2 times the mean value of the net annuity premium released by an annuitant's death. He assumed that the distribution of these net annuity premiums was

$$B(x) = \frac{c^c}{\Gamma(c)} e^{-cx} x^{c-1}, \quad c > 0, \quad 0 < x < \infty, \tag{5.2}$$

so that

$$\mathcal{E}x = 1 \quad \text{and} \quad \mathcal{U}x = \frac{1}{c},$$

giving

$$\hat{c} = \frac{1}{2.2}.$$

We thus have to determine κ and η from

$$\frac{c^c}{\Gamma(c)} \int^{\infty} e^{-(\kappa+c)x} x^{c-1} \, dx = \left(\frac{c}{\kappa + 1}\right)^c = 1 - \kappa(1 - \eta) \tag{5.3}$$

and

$$\psi(z) \sim e^{-\kappa z}, \tag{5.4}$$

where z is the amount of risk reserve in units equal to the mean net annuity premium released on death, $c = 1/2.2$, and $\psi(z)$ is chosen as 0.0025, say. For this set of parameters we have the following table of η-values (Table 5.1).

Table 5.1

z	100η
10	31.38
100	4.20
1,000	0.44
10,000	0.08

The risk loading rapidly becomes insignificant once the risk reserve exceeds one hundred times the mean sum at risk. We refer the reader to Segerdahl (1955; cited in Chapter 4) for the moments of the time to ruin (given that it is eventually certain) in this case of negative Y-values.

Another method of determining η_1 would be to require that the aggregate claims during a suitably chosen period should, with probability equal to 0.999 (say) be less than the current risk reserve plus the premiums for that period. This implies that η_1 is to be calculated from

$$F(\kappa_1 + \eta_1 + x) = 0.999, \tag{5.5}$$

where x is the risk reserve. Any of the methods of calculating $F(\cdot)$ described in Chapter 2 could be used in the foregoing relation. Because this method requires only the existence of a positive risk reserve at the end of the chosen period and ignores the possibility that ruin may have occurred before the end of that period the risk loadings calculated from (5.5) and (2.33) will be less than those produced by the first method for given x and t.

We mention that the idea expressed by (5.5) was first broached by Wittstein in 1885 (Wagner, 1898), who considered that the risk loading on the premiums of each of N identical policies of a given portfolio should be equal to one-Nth of the standard deviation of the (present value of the) aggregate outgo.

REINSURANCE: INTRODUCTION

Suppose that two risk businesses have equal risk reserves and load their risk premiums by the same percentage. Then it seems obvious that in general the company whose $F(\cdot)$ has the longer tail will have the larger probability of being ruined. This suggests that a risk business should give careful attention to maintaining its existing distribution $F(\cdot)$ by rejecting or limiting the potential claim outgo of new contracts or might even consider how to reduce its probability of ruin during a shorter or longer period by ceding a portion of its risk portfolio to another risk business. The former problem is known in the older literature as that of "maximum retention," whereas the more general approach is called reinsurance. In this chapter it is assumed that the risk business can always find a reinsurer willing to accept—for a price*—any portion of its portfolio of risks. In the next chapter we consider the reciprocal problems of two or more insurance companies who desire to exchange portions of their risk portfolios.

There are three common types of reinsurance contract that we shall consider:

1. Quota (or proportional) reinsurance in which a fixed proportion of every individual sum insured (or sum-at-risk) is reinsured.

2. Excess-loss (or surplus) reinsurance in which individual sums insured (or sums-at-risk) in excess of a fixed amount M are reinsured. (This is mathematically equivalent to a variable quota reinsurance.)

3. Stop-loss ("excess-loss" in the earlier literature) reinsurance in which the aggregate claims of any year in excess of a "priority" $X = c$ are paid by the reinsurer.

---

* If the loading included in the reinsurance premium exceeds the corresponding portion of the loading imposed on the net premium for the whole risk, the principal insurer (cedent) must decide whether it will accept this sure loss instead of rejecting the coverage of the corresponding portion of the risk. Hereafter we shall assume that the principal company's loading rate always exceeds that of the reinsurer.

MAXIMUM RETENTION

Much of the published work on the calculation of the maximum sum insured a company should be prepared to assume relates to life insurance sums at risk and dates back to a time when the within-portfolio classification of contracts (see Chapter 2) was the preferred mathematical model. The criterion for limiting individual claim amounts was thus based on the probability of the current risk reserve plus the prospective premium income being in excess of the aggregate claim outgo by the end of the next year or possibly by the end of the potential future duration of the individual long-term life insurance contracts of the portfolio. The typical precedure was to use the Normal approximation to calculate the foregoing probability both before and after the retention limit had been imposed. If the increase thus produced in the probability of nonruin were deemed insufficient, a lower limit of retention could be imposed.

The earliest approach, however, was to ignore the risk reserve and the prospective risk loadings and to assume that the aggregate claim d.f. of the existing portfolio of contracts was satisfactory. Thinking in terms of the Normal distribution, the problem was to determine the maximum sum insured under a single new independent contract which would maintain the existing value of the coefficient of variation of the aggregate outgo. An interesting result was that under fairly general assumptions this led to a maximum retention of approximately twice the mean sum-at-risk of the existing portfolio (Berger, 1925, cited in Chapter 2). Meidell (1953) has considered the application of this rule when the individual sums-at-risk are distributed according to the Pareto law (2.32).

Later, the risk reserve and prospective risk loadings were included in the calculation (Berger, 1925; Dubourdieu, 1952). Write R for the current amount of the risk reserve and E for the expected aggregate of future risk loadings from the existing portfolio of contracts—the expectation being a fixed amount if the premiums are paid annually in advance and only one future year is under consideration. Using the cumulant notation $\kappa_j (j = 1, 2)$ for the mean and variance of the existing portfolio and $s^j k_j (j = 1, 2)$ for the corresponding values for a single new contract with sum insured s and expected risk loading $s\eta_1$, the probability of a state of nonruin at the end of the chosen period (namely, at the end of one year or at the eventual termination of all contracts) is presently

$$\Phi\left(\frac{R + E}{\sqrt{\kappa_2}}\right) \quad \text{and will become} \quad \Phi\left(\frac{R + E + s\eta_1}{\sqrt{\kappa_2 + s^2 k_2}}\right),$$

when the new contract has been added to the portfolio. The curve of

$(R + E + s\eta_1)/(\kappa_2 + s^2 k_2)^{1/2}$ as a function of s attains a maximum when

$$s = \frac{\eta_1}{k_2} \cdot \frac{\kappa_2}{R + E} \qquad (5.6)$$

has a value equal to $(R + E)/\sqrt{\kappa_2}$ when

$$s = 2\frac{\eta_1}{k_2} \cdot \frac{\kappa_2}{R + E} \cdot \left(1 - \frac{\eta_1}{\sqrt{k_2}} \cdot \frac{\sqrt{\kappa_2}}{R + E}\right)^{-1} \qquad (5.7)$$

and tends asymptotically to a value $\eta_1/\sqrt{k_2}$. Notice that when $R + E$ is large relative to $\sqrt{\kappa_2}$ s determined from (5.7) is approximately twice the value obtained from (5.6). Deitz (1954) has generalized the preceding arguments to the case in which n different new independent contracts are to be written. An alternative procedure is to maximize $R + E + s\eta_1 - \varepsilon(\kappa_2 + s^2 k_2)^{1/2}$, where $\varepsilon > 0$ is a suitably chosen "coefficient of security" and this produces

$$s = \frac{\eta_1}{k_2} \cdot \frac{\sqrt{\kappa_2}}{(\varepsilon^2 - \eta_1^2/k_2)^{1/2}} \quad \text{with} \quad \frac{\eta_1^2}{k_2} < \varepsilon^2, \qquad (5.8)$$

which is independent of $R + E$.

The foregoing results are based on the idea that a prospective policy-holder's sum insured is to be limited in amount according to suitable rules. With the general availability of reinsurance contracts from other insurance companies, the emphasis is now on a review of the whole portfolio of risks to determine whether a portion of some or all of them should be reinsured. Using the model that leads to a d.f. of X given by (2.33), we may seek a maximum retained sum-at-risk M determined from α, the probability of exhausting the expected risk reserve before the end of t years, namely

$$F(R + \eta\kappa_1, t) = 1 - \alpha, \qquad (5.9)$$

where

$$\kappa_1 = \lambda t \left\{ \int_0^M y \, dP(y) + M[1 - P(M)] \right\} = \lambda t \int_0^M [1 - P(y)] \, dy \qquad (5.10)$$

and

$$\kappa_2 = \lambda t \left\{ \int_0^M y^2 \, dP(y) + M^2[1 - P(M)] \right\}$$

$$= 2\lambda t \int_0^M y[1 - P(y)] \, dy. \qquad (5.11)$$

Shirasugi (1954) has illustrated (5.9) numerically by using (2.42) without the last term and assuming the binomial with $q = 0.01$ instead of Poisson for $p_n(1)$; G. Berger (1964) has considered the case in which $P(y)$ is a mixture of two negative exponentials and $F(\cdot)$ is Normal; and Cacciafesta (1962) takes advantage of the assumed normality of $F(\cdot)$ to derive an M-value which minimizes the aggregate variance.

Following Pentikäinen (1952), we may derive an approximation for M as follows: write the standardized value of $R + (1 + \eta)\kappa_1$ equal to z; then from (5.9)

$$R + (1 + \eta)\kappa_1 - \kappa_1 = z\sqrt{\overline{\kappa_2}}$$

or

$$\frac{(R + \eta\kappa_1)^2}{\kappa_1} \equiv g^2(\alpha, M) = z^2 \frac{\kappa_2}{\kappa_1}, \qquad (5.12)$$

where z is a function of α. But, from (5.10) and (5.11),

$$\frac{\kappa_2}{\kappa_1} = 2 \frac{\displaystyle\int_0^M x[1 - P(x)]\,dx}{\displaystyle\int_0^M [1 - P(x)]\,dx}$$

$$\simeq 2 \frac{\displaystyle\int_0^M x[1 - P(M)x/M]\,dx}{\displaystyle\int_0^M [1 - P(M)x/M]\,dx}$$

$$= M \frac{1 - 2P(M)/3}{1 - P(M)/2} \sim \tfrac{2}{3}M. \qquad (5.13)$$

if M is large and $P(M)$ approximately unity. Relation 5.12 thus becomes

$$M \sim \frac{3}{2z^2} \cdot \frac{(R + \eta\kappa_1)^2}{\kappa_1}$$

and, if $F(\cdot)$ were Normal and α were chosen equal to 0.10, we would obtain $z = 2.326$ and

$$M \sim 0.277 \frac{(R + \eta\kappa_1)^2}{\kappa_1}, \qquad (5.14)$$

which is a simple formula to determine M, given R, κ_1, and η. Thus, for example, a risk business with risk reserves equal to 100 times its mean sum-at-risk (before reinsurance of surplus risks) and a risk loading of 5% of all net premiums would need to set an approximate limit of

$$M = 0.277 \frac{(100 + \kappa_1/20)^2}{\kappa_1}$$

on its retained sum-at-risk if it were to have a 99% probability of nonruin when it reached the end of an accounting period corresponding to κ_1.

Suppose that this period were to be equated to that during which 4000 claims were expected; then

$$M = 0.277 \frac{300^2}{4000} = 6.2.$$

For this to be valid the mean sum-at-risk after reinsurance of risks in excess of 6.2 times the original mean sum-at-risk should still be approximately unity.

Pentikäinen (*loc. cit*) used (2.49) with the term in $A_3(\cdot)$ suppressed to calculate $g^2(\cdot)$ in (5.12) as a function of M with $\alpha = 0.01$ for 10 different sample distributions $F(\cdot)$. The g-values for these distributions satisfied

$$g^2(0.01, M) < 3.61M = \frac{M}{0.277}, \tag{5.15}$$

namely (5.14). Further developments of these ideas are found in Pentikäinen (1954).

Instead of using the probability of ruin (at least once) by the end of a given period, we may determine M by requiring the probability of *ultimate* ruin to assume a given small value. In all the published literature on this question numerical solutions have been based on the following approximate (asymptotic) equivalence obtained from (4.64), namely,

$$v(x, \infty) \equiv \psi(x) \simeq e^{-\kappa x}, \tag{5.16}$$

where κ is given by (4.62); that is,

$$1 + \kappa(p_1 + \eta_1) = \int_0^\infty e^{\kappa y}\, dP(y), \tag{5.17}$$

when $p_n(t)$ is Poisson and with the left-hand side of this relation replaced by

$$1 + \left(\frac{t}{c}\right)^{-1} \{1 - \exp[-(p_1 + \eta_1)\, \kappa\, t/c]\}, \tag{5.18}$$

when $p_n(t)$ is negative binomial with the form (3.35) with $m = c$ so that its mean is t and its variance $t(1 + t/c)$.* The usual choice of p_1 as the unit of sum-at-risk requires modification, since, after reinsurance of sums in excess of M, a "lump" of probability $1 - P(M)$ is ascribed to the value $Y = M$ and the mean of Y has changed. This question is considered in detail by Segerdahl (1948), who also develops criteria for comparing two $P(\cdot)$ and investigates the case in which $P(\cdot)$ is negative exponential in some detail.

Two of the numerical studies were based on actual insurance company data for $P(y)$. In the first, which utilized 5.17 and the Thule insurance company's

* When $c \to \infty$ (5.18) reduces to the left of (5.17).

portfolio of Ordinary Life sums-at-risk during the years 1929–1931, Hultmann (1942) in his Table 8 provided values of M (Table 5.2) as multiples of the mean sum-at-risk prior to reinsurance (namely, Kr. 4061) when $\psi(x) = 10^{-3}$ in (5.16), x also being expressed in the same monetary unit:

Table 5.2

η	x		
	100	500	1000
0.05	2.4	27.6	65.0
0.10	5.6	62.0	
0.15	9.9		
0.20	14.5		
0.25	18.8		

The corresponding table of M-values given by Ammeter (1954) was derived by using a distribution $P(\cdot)$ based on "langjährige Erfahrungen" (possibly those mentioned below). These values of M shown in Table 5.3 were based on

Table 5.3

η	x			
	50	100	200 ($c = 100$)	200 ($c = \infty$)
0.2	3	18	39	67
0.3	10	36	75	108
0.4	17	54	110	145
0.5	23	66	140	174
0.6	27	76	163	195
0.7	32	86	183	214
0.8	36	95	200	231

$\psi(x) = 10^{-2}$ and, for $x = 200$, (5.18) was also used. Earlier, Ammeter (1946) had utilized the Swiss Life and Annuity Company's portfolio of insurance sums-at-risk to produce a graduated $P(\cdot)$ in which a negative exponential was used for sums in excess of 10,000 Fr. He considered how the effective risk loading was reduced by having to pay a loading on the reinsured portion of the net premium—a matter referred to earlier by Hultmann (1942).

Two different approximations have been suggested to simplify the iterative calculations required by (5.17). In the first Thépaut (1953) writes, in effect,

$$\kappa(p_1 + \eta_1) = \int_0^\infty (e^{\kappa y} - 1)\, dP(y) \simeq \int_0^\infty \left\{ \kappa y + \frac{(\kappa y)^2}{2} \left[1 - \frac{\kappa y}{3} \right]^{-1} \right\} dP(y);$$

that is,

$$\eta_1 \equiv p_1\eta \simeq \frac{\kappa}{2} \int_0^\infty y^2 \left[1 - \frac{\kappa y}{3} \right]^{-1} dP(y). \qquad (5.19)$$

Now, when there is a mass of probability of size $1 - P(M)$ at the point $y = M$, this approximate equation becomes

$$\frac{2\eta}{\kappa} \simeq \left\{ \int_0^M y^2 \left(1 - \frac{\kappa y}{3} \right)^{-1} dP(y) + M^2 \left(1 - \frac{\kappa M}{3} \right)^{-1} [1 - P(M)] \right\} p_1^{-1}$$

$$< \left(1 - \frac{\kappa M}{3} \right)^{-1} \frac{\displaystyle\int_0^M y^2 \, dP(y) + M^2[1 - P(M)]}{\displaystyle\int_0^M y \, dP(y) + M[1 - P(M)]} .$$

Writing the quotient last written as ωM, we finally obtain

$$M > \frac{6\eta}{\kappa(3\omega + 2\eta)},$$

which is written by Thépaut in the form

$$\frac{M}{x} > \frac{6\eta}{\kappa x(3\omega + 2\eta)}, \qquad (5.20)$$

where x is the risk reserve expressed in the same monetary units as p_1. A very simple example may be taken from Thépaut (1957). If $P(y)$, $0 < y < 1$, is the d.f. of the proportion y of the sum insured that becomes payable and, in fact, no proportions other than unity can occur and all sums insured are equal, then $\omega = 1$ and, with $\eta = 0.1$ and $\kappa x = 13.82$ (corresponding to $e^{-\kappa x} = 10^{-6}$)

$$\frac{M}{x} > \frac{6 \times 0.1}{13.82(3 + 0.2)} = 0.0136.$$

Hence such a business can accept a uniform maximum sum insured of at least 1.36% of its risk reserve x. In this second article Thépaut provides a series of examples of the use of (5.20) based on proportionate losses occurring to classes of French maritime vessels during the years 1950–1954.

The second approximation, by Lambert (1960), proceeds from (5.17) as follows: given η_1 and with κ determined from (5.16), the equation for the maximum retention M is

$$1 + \kappa(p_1 + \eta_1) = \int_0^M e^{\kappa y} \, dP(y) + e^{\kappa M}[1 - P(M)]$$

$$= \sum_{j=0}^\infty \frac{(\kappa M)^j}{j!} \left\{ \int_0^M \left(\frac{y}{M} \right)^j dP(y) + [1 - P(M)] \right\}; \qquad (5.21)$$

that is,

$$\eta_1 = \sum_{j=2}^{\infty} \frac{(\kappa M)^{j-1}}{j!} M\left[\int_0^M \left(\frac{y}{M}\right)^j dP(y) + 1 - P(M)\right]$$

$$\simeq \frac{\kappa M}{2} \tau_2(M) + \frac{(\kappa M)^2}{6} \tau_3(M), \qquad (5.22)$$

where

$$\tau_j(M) \equiv M\left[\int_0^M \left(\frac{y}{M}\right)^j dP(y) + 1 - P(M)\right], \qquad j = 2, 3.$$

Relation 5.22 may be relatively easy to evaluate when $P(\cdot)$ is given empirically, since we need to calculate only the first three moments about zero of the retained portion of the distribution.

We mention that Lambert (1960) also showed that for a given risk reserve and (approximate) probability of ultimate ruin the loadings η_1 obtained for portfolios modified from an original portfolio by refusing risks in a manner analogous to that achieved by stop-loss, excess-loss, and proportionate reinsurance, respectively, are in increasing order of size. Of course, this is not necessarily the case when reinsurance is actually effected because the differing costs of various types of reinsurance must then be considered.

REINSURANCE BASED ON INDIVIDUAL RISKS

The idea of determining the optimal reinsurance quota for each individual contract apparently originated with Medolaghi (1909). Write s_i for the sum-at-risk of contract $i(i = 1, 2, \ldots, N)$, σ_i^2 for that contract's variance, η_i for the loading loss suffered by reinsuring the whole of the contract, and α_i for the proportion (quota) to be reinsured. It is proposed to minimize the aggregate variance of the portfolio remaining after reinsurance, subject to a fixed aggregate loading loss A on the reinsured portions of the contracts. It is assumed that the N contracts are uncorrelated; the contrary case is treated by Finetti (1940).

Since

$$A = \sum_{i=1}^{N} \alpha_i \eta_i,$$

we minimize

$$\sum_{i=1}^{N} (1 - \alpha_i)^2 \sigma_i^2 - \lambda\left(A - \sum_{i=1}^{N} \alpha_i \eta_i\right), \qquad (5.23)$$

where λ is an undetermined coefficient, for variation in $\alpha_i(i = 1, 2, \ldots, N)$. Differentiating (5.23) with respect to α_i and equating the result to zero, we obtain

$$\alpha_i = 1 - \frac{\lambda}{2} \cdot \frac{\eta_i}{\sigma_i^2}$$

and substituting this into A we have

$$\frac{\lambda}{2} = \frac{\sum\limits_{i=1}^{N} \eta_i - A}{\sum\limits_{i=1}^{N} \eta_i^2 / \sigma_i^2} .$$

Hence

$$\alpha_i = 1 - \frac{\sum\limits_{i=1}^{N} \eta_i - A}{\sum\limits_{i=1}^{N} \eta_i^2 / \sigma_i^2} \cdot \frac{\eta_i}{\sigma_i^2} , \qquad (5.24)$$

and it is easily verified that the resulting value of the aggregate variance, namely,

$$\frac{\left(\sum\limits_{i=1}^{N} \eta_i - A \right)^2}{\sum\limits_{i=1}^{N} \eta_i^2 / \sigma_i^2} \qquad (5.25)$$

is a minimum.

The optimal quota provided by (5.24), however, is not necessarily positive, and we have to investigate whether the result of replacing any negative values of the right-hand side of (5.24) with zero is to leave the aggregate variance a minimum subject to $\alpha_i \geq 0$. Suppose that the contracts have been numbered in such a way that the corresponding quantities σ_i^2 / η_i ($i = 1, 2, \ldots, N$) are in increasing order of magnitude. Let v be the integer such that

$$\frac{\sigma_v^2}{\eta_{v-1}} < \frac{\sum\limits_{i=v}^{N} \eta_i - A}{\sum\limits_{i=v}^{N} \eta_i^2 / \sigma_i^2} \leq \frac{\sigma_v^2}{\eta_v} ; \qquad (5.26)$$

then it can be shown (Medolaghi, *loc. cit.*) that the quota system

$$\alpha_i = \begin{cases} 0, & i = 1, 2, \ldots, v - 1, \\ 1 - \dfrac{\sum\limits_{i=v}^{N} \eta_i - A}{\sum\limits_{i=v}^{N} \eta_i^2 / \sigma_i^2} \cdot \dfrac{\eta_i}{\sigma_i^2}, & i = v, v + 1, \ldots, N, \end{cases} \qquad (5.27)$$

results in

$$\sum\limits_{i=1}^{v-1} \sigma_i^2 + \sum\limits_{i=v}^{N} (1 - \alpha_i)^2 \sigma_i^2 < \sum\limits_{i=1}^{N} (1 - \alpha_i')^2 \sigma_i^2, \qquad (5.28)$$

where

$$\alpha_i' = \alpha_i + \beta_i$$

and the β_i are subject to $\sum_{i=1}^{N} \beta_i \eta_i = 0$.

We note that α_i in (5.24) is a constant only when η_i is proportional to $\sigma_i{}^2$, and the usual quota reinsurance is then optimal in the sense used above. This type of risk loading conforms to the requirements of quadratic utility (Chapter 6).

On the other hand, reinsurance of the excess over a fixed maximum retention M (excess-loss reinsurance) is optimal only when η_i is proportional to $\sigma_i{}^2/s_i$.

A similar type of analysis, but ignoring the loadings included in the individual policy premiums, was made by Ottaviani (1952). He envisaged n insurance companies all effecting quota reinsurances with the same reinsurance company. Policy j in company i ($j = 1, 2, \ldots, n_i$) has a normally distributed outgo with expectation (net premium) ξ_{ij} and variance $\sigma_{ij}{}^2$ and a proportion α_{ij} of it is to be reinsured. If company i is to have a fixed net reinsurance premium of $\alpha_i \sum_{j=1}^{n_i} \xi_{ij}$ ($i = 1, 2, \ldots, n$), it is shown that equalization of the probabilities of ruin of the n companies and the reinsurer by the end of the year, and minimization thereof, implies minimization of

$$\left(\sum_{i=1}^{n} \sum_{j=1}^{n_i} \alpha_{ij}{}^2 \sigma_{ij}{}^2 \right)^{1/2} + \sum_{i=1}^{n} \left[\sum_{j=1}^{n_i} (1 - \alpha_{ij})^2 \sigma_{ij}{}^2 \right]^{1/2}.$$

The optimal quota for reinsurance by company i on policy j is then

$$\alpha_{ij} = h_i + \frac{k_i \xi_{ij}}{\sigma_{ij}{}^2},$$

where

$$h_i{}^{-1} = 1 + \left[\sum_{j=1}^{n_i} (1 - \alpha_{ij})^2 \sigma_{ij}{}^2 \right]^{1/2} \left(\sum_{i=1}^{n} \sum_{j=1}^{n_i} \alpha_{ij}{}^2 \sigma_{ij}{}^2 \right)^{1/2}$$

and

$$k_i = \left[h_i \sum_{j=1}^{n_i} \xi_{ij} - \sum_{j=1}^{n} (1 - \alpha_{ij}) \xi_{ij} \right] \left(\sum_{j=1}^{n_i} \frac{\xi_{ij}{}^2}{\sigma_{ij}{}^2} \right)^{-1}.$$

The idea of a fixed maximum retention of M carries over directly into the reinsurance area. The only difference is that instead of losing a certain portion of the loading η_i corresponding to the excess sum-at-risk $s_i - M > 0$ this excess is reinsured at a cost that is possibly less than the risk loading received by the principal insurer. If this is so, the theory adumbrated under maximum retention is applicable also to this excess-loss reinsurance.

A modified form of excess-loss reinsurance has been proposed by Benktander (1964). It is based on the payment of a premium rate fixed in advance applied a posteriori to the number of claims that exceed an agreed amount y_0. Suppose, for example, that the excess over M of every claim that exceeds $M > y_0$ is to be reinsured and that the net premium rate has been fixed at

$$\frac{\text{expected number of claims with } Y > M}{\text{expected number of claims with } Y > y_0}$$

$$\times \text{ (mean value of excess claim in excess of } M)$$

$$= \frac{1 - P(M)}{1 - P(y_0)} \times [1 - P(M)]^{-1} \int_M^\infty (y - M)\, dP(y) = p_{1B} \quad \text{(say)}$$

if numbers of claims and their sizes are independent. Note that p_{1B} depends only on $P(\cdot)$ and not on the distribution of the number of claims.

Write L for the random variable denoting the number of claims between y_0 and M, and N for that denoting the number of claims in excess of M. These random variables are independent so that $\mathcal{E}\{L(N - \mathcal{E}N)\} = 0$. The reinsurance premium is $(L + N)$ times p_{1B} and its expected value is

$$\mathcal{E}[(L + N)p_{1B}] = \mathcal{E}(L + N)\left\{ \frac{\mathcal{E}(N)}{\mathcal{E}(L + N)}\, [1 - P(M)]^{-1} \int_M^\infty (y - M)\, dP(y) \right\}$$

$$= \text{ expected aggregate of excess claims in excess of } M.$$

The net premium is thus unbiased, and if we put $\bar{Z}$ for the average of N positive values of $Y - M$ the foregoing relation may be written

$$\mathcal{E}[(L + N)p_{1B}] = \mathcal{E}N\mathcal{E}\bar{Z} = \mathcal{E}(N\bar{Z}).$$

The reinsurer's gain being $(L + N)p_{1B} - N\bar{Z}$ with zero mean its variance is

$$\mathcal{E}\{(L + N)p_{1B} - N\bar{Z}\}^2 = \mathcal{E}[\{(L + N)p_{1B} - \mathcal{E}N\mathcal{E}\bar{Z}\} - (N\bar{Z} - \mathcal{E}N\mathcal{E}\bar{Z})]^2$$

$$= \mathcal{E}\{(L + N)p_{1B} - \mathcal{E}N\mathcal{E}\bar{Z}\}^2 + \mathcal{E}(N\bar{Z} - \mathcal{E}N\mathcal{E}\bar{Z})^2$$

$$\quad - 2\mathcal{E}[\{(L + N)p_{1B} - \mathcal{E}N\mathcal{E}\bar{Z}\}(N\bar{Z} - \mathcal{E}N\mathcal{E}\bar{Z})]$$

$$= \mathcal{E}(L + N)^2 p_{1B}^2 - \{\mathcal{E}(L + N)\}^2 p_{1B}^2 + \mho(N\bar{Z})$$

$$\quad - 2p_{1B}\mathcal{E}[N(N\bar{Z} - \mathcal{E}N\mathcal{E}\bar{Z})]$$

$$= p_{1B}^2 \mho(L + N) + \mho(N\bar{Z})$$

$$\quad - 2p_{1B}\{\mathcal{E}N^2 - (\mathcal{E}N)^2\}\mathcal{E}\bar{Z}.$$

Benktander (loc. cit.) assumes that L and N are independently Poisson distributed so that $\mathcal{E}(L + N) = \mho(L + N)$ and $\mathcal{E}N = \mho N$. This produces

$$\mho\{(L + N)p_{1B}\} = \mho(N\bar{Z}) - p_{1B}\mathcal{E}N\mathcal{E}\bar{Z} = \mho(N\bar{Z}) - \frac{(\mathcal{E}N\mathcal{E}\bar{Z})^2}{\mathcal{E}(L + N)}$$

and means that the reinsurer does not absorb the whole of the variance of the excess claims, namely, $\mho(N\bar{Z})$. In the article cited a numerical study is made of the special case in which $P(y)$ is Pareto, namely, of the form (2.32), which implies that p_{1B} is independent of M and thus insensitive to changes in the value of money.

A recent paper (Benktander & Ohlin, 1967) considers the effect of a double layer of reinsurance on a fire portfolio. The upper layer consists of excess-loss (here called "surplus") reinsurance of that part of each sum insured (here called the "probable maximum loss," PML) in excess of M; the corresponding net premium is

$$\rho_M = \int_M^\infty (y - M)\, dP(y) \int_0^1 x\, dK_y(x),$$

where $K(\cdot)$ is the d.f. of the proportion of the building actually consumed by fire, made by the authors to depend on the size of the building (PML). The lower layer of reinsurance is for actual losses in excess of m and has a net premium

$$\rho_M(m) = \int_m^M (x - m)\, dH(x),$$

where

$$1 - H(x) = \int_x^M \left[1 - K_y\left(\frac{x}{y}\right)\right] dP(y) + \int_M^\infty \left[1 - K_y\left(\frac{x}{M}\right)\right] dP(y).$$

These premiums are evaluated for two special cases, in both of which $P(\cdot)$ is given by (2.32) with $\alpha = 2$. In the first case

$$K_y(x) = K(x) = x, \qquad 0 < x < 1,$$

and in the second

$$K_y(x) = \frac{x^{1-\beta} - (1/y)^{1-\beta}}{1 - \beta}, \qquad 0 < x < 1.$$

FORMS OF STOP-LOSS REINSURANCE

We mentioned in Chapter 3 that the large standard error involved in the estimate of κ_{1c}, the mean value of the aggregate outgo X in excess of a priority c, has inhibited the development of stop-loss reinsurance under which the reinsurer undertakes responsibility for the excess (if any) of aggregate claims over a given priority.* This is why several authors have proposed modified forms of stop-loss reinsurance.

One suggestion (Thépaut, 1950) is to reinsure the excess of each of the company's n largest claims over the amount of that nth claim, where n (say, 20) is chosen in advance but the reinsurance premium is paid in arrear only after the actual size of the nth largest claim is known. Thépaut proposed that this reinsurance treaty of the "excédent du coût moyen relatif"

---

* Finetti (1940) calls such reinsurance "eccessivamente aleatorio"; Richard (1949) considers it "impossible ou illusoire."

(ECOMOR) should have its *rate* of reinsurance premium fixed in advance by means of a Pareto distribution chosen to represent the upper tail of the time-invariant distribution of individual claims.

Thus suppose $y_{(n)}$ is the observed value of the random variable $Y_{(n)}$ corresponding to the nth largest claim and let the d.f. of Y, the individual claim size, be

$$P(y) = 1 - \left(\frac{y}{y_0}\right)^{-\alpha}, \qquad y_0 < y < \infty. \tag{5.29}$$

Given that a claim has a value in excess of $y_{(n)}$, its mean excess value is

$$p_{1y} = \frac{\displaystyle\int_{yn}^{\infty} (y - y_{(n)})\, dP(y)}{\displaystyle\int_{y(n)}^{\infty} dP(y)} = \frac{y_{(n)}}{\alpha - 1}, \tag{5.30}$$

and the aggregate net premium for reinsurance of the sum of the excesses of each of n claims over $y_{(n)}$ is $(n-1)p_{1y}$, which is independent of y_0. This latter property is valuable in a period of rapid inflation, provided the Pareto parameter α does not change with the purchasing power of money. Thépaut (1950) found that the α-values for French fire insurance claims during the immediate pre- and postwar years were around 1.5 to 2.0. This implies an infinite variance for $P(\cdot)$, and it is a feature of Thépaut's proposal that there should be a uniform ceiling on each of the n largest claims above which the principal insurer would again be responsible. We have ignored this ceiling in what follows.

The reinsurer under an ECOMOR treaty has thus undertaken to pay

$$\sum_{j=1}^{n} Y_{(j)} - n Y_{(n)}.$$

His expected payment is then

$$\rho_n = \mathcal{E}\left[\sum_{j=1}^{n} Y_{(j)} - n Y_{(n)}\right] \tag{5.31}$$

and the variance of his payments is

$$V_n = \mathcal{E}\left(\sum_{j=1}^{n-1} Y_{(j)}^2 + 2\sum_{i=1}^{n-2}\sum_{j=i+1}^{n-1} Y_{(i)}Y_{(j)} + (n-1)^2 Y_{(n)}^2 \right.$$
$$\left. - 2(n-1)Y_{(n)}\sum_{j=1}^{n-1} Y_{(j)}\right) - \rho_n^2. \tag{5.32}$$

Now, if the d.f. of the aggregate amount of claim outgo when expected claims number t is given by (2.33), the density function of the nth largest claim size $Y_{(n)}$ is

$$\sum_{k=n}^{\infty} e^{-t}\frac{t^k}{k!}\binom{k}{k-n}[P(y)]^{k-n}[1 - P(y)]^{n-1} n\, dP(y),$$

since $k - n$ of the k claims must have amounts less than y, $n - 1$ of the remainder must be greater than y, and one claim (selected from among n) must have a claim size of exactly y. The foregoing density easily reduces to

$$e^{-t[1-P(y)]} \frac{t^n}{(n - 1)!} [1 - P(y)]^{n-1} dP(y), \tag{5.33}$$

as given by Ammeter (1964). Further, by an analogous argument the joint density function of $Y_{(i)}$ and $Y_{(j)}$ $(i < j)$ is $(y_0 < y < z < \infty)$.

$$\sum_{k=j}^{\infty} e^{-t} \frac{t^k}{k!} \binom{k}{k - j} [P(y)]^{k-j} j \, dP(y) \binom{j - 1}{j - i - 1}$$

$$\times [P(z) - P(y)]^{j-i-1} i \, dP(z)[1 - P(z)]^{i-1}$$

$$= e^{-t[1-P(y)]} \frac{t^j}{(j - i - 1)! \, (i - 1)!} [P(z) - P(y)]^{j-i-1}$$

$$\times [1 - P(z)]^{i-1} dP(y) \, dP(z). \tag{5.34}$$

When $P(\cdot)$ assumes the Pareto form (5.29) with $y_0 = 1$ for convenience, (5.33) provides

$$\mathcal{E} Y_{(j)}^{\nu} = \frac{t^j}{(j - 1)!} \int_1^{\infty} e^{-ty^{-\alpha}} y^{\nu - \overline{\alpha j} - 1} \alpha y^{-\alpha - 1} \, dy = \frac{t^{\nu/\alpha}}{(j - 1)!} \int_0^t e^{-x} x^{j - \nu/\alpha - 1} \, dx$$

$$= \frac{t^{\nu/\alpha}}{(j - 1)!} \gamma(j - \nu/\alpha, t), \qquad \nu = 1, 2, 3, \ldots,$$

in incomplete gamma function notation (Magnus et al., 1966).

Similarly (5.34) yields

$$\mathcal{E} Y_{(i)} Y_{(j)} = \frac{t^j}{(j - i - 1)! \, (i - 1)!} \int_1^{\infty} z^{1 - \overline{\alpha i} - 1} \alpha z^{-\alpha - 1} \, dz$$

$$\times \int_1^z e^{-ty^{-\alpha}} (y^{-\alpha} - z^{-\alpha})^{j-i-1} y \alpha y^{-\alpha - 1} \, dy$$

$$= \frac{t^{2/\alpha}}{(j - i - 1)! \, (i - 1)!} \int_0^t v^{i - 1/\alpha - 1} \, dv \int_v^t x^{-1/\alpha} (x - v)^{j-i-1} e^{-x} \, dx$$

$$= \frac{t^{2/\alpha}}{(j - i - 1)! \, (i - 1)!} \int_0^t x^{-1/\alpha} e^{-x} \, dx \int_0^x v^{i - 1/\alpha} (x - v)^{j-i-1} \, dv$$

$$= \frac{t^{2/\alpha}}{(j - i - 1)! \, (i - 1)!} \sum_{k=0}^{j-i-1} (-1)^k \binom{j - i - 1}{k}$$

$$\times \int_0^t x^{j-i-1-k-1/\alpha} e^{-x} \, dx \int_0^x v^{k+i-1/\alpha - 1} \, dv$$

$$= \frac{t^{2/\alpha} \gamma(j - 2/\alpha, t)}{(j - i - 1)! \, (i - 1)!} \sum_{k=0}^{j-i-1} (-1)^k \binom{j - i - 1}{k} \frac{1}{k + i - 1/\alpha}.$$

Now it is easy to prove that

$$\sum_{j=1}^{n-1} \frac{\Gamma(j - 1/\alpha)}{\Gamma(j)} = \frac{\alpha}{\alpha - 1} \frac{\Gamma(n - 1/\alpha)}{\Gamma(n - 1)},$$

so that when t is large and $\gamma(j - 1/\alpha, t) \sim \Gamma(j - 1/\alpha)$

$$p_n \sim t^{1/\alpha} \sum_{j=1}^{n-1} \frac{\Gamma(j - 1/\alpha)}{\Gamma(j)} - (n - 1)t^{1/\alpha} \frac{\Gamma(n - 1/\alpha)}{\Gamma(n)}$$

$$= \frac{(n - 1)t^{1/\alpha}}{\alpha - 1} \frac{\Gamma(n - 1/\alpha)}{\Gamma(n)} = (n - 1) \frac{\mathcal{E}Y_{(n)}}{\alpha - 1},$$

in agreement with Ammeter (1964). The (posterior) net premium $(n - 1)p_{1y}$ is thus consistent but biased.

Turning now to V_n, we note that $(i < j)$

$$\mathcal{E}\left\{ Y_{(j)} \sum_{i=1}^{j-1} Y_{(i)} \right\} = t^{2/\alpha} \gamma(j - 2/\alpha, t) \sum_{i=1}^{j-1} \frac{1}{(i - 1)!} \sum_{k=0}^{j-i-1} (-1)^k$$

$$\times \frac{1}{k!(j - i - 1 - k)!} \cdot \frac{1}{k + i - 1/\alpha}$$

$$= t^{2/\alpha} \gamma(j - 2/\alpha, t) \sum_{i=1}^{j-1} \frac{(-1)^i}{(i - 1)!} \sum_{l=i}^{j-1} (-1)^l$$

$$\times \frac{1}{(l - i)!(j - l - 1)!} \cdot \frac{1}{l - 1/\alpha}$$

$$= t^{2/\alpha} \gamma(j - 2/\alpha, t) \sum_{l=1}^{j-1} (-1)^{l+1}$$

$$\times \frac{1}{(l - 1)!(j - l - 1)!} \cdot \frac{1}{l - 1/\alpha} \sum_{i=1}^{l} (-1)^{i-1} \binom{l - 1}{i - 1}$$

$$= 0, \quad \text{since } \sum_{m=0}^{l-1} (-1)^m \binom{l - 1}{m} = (1 - 1)^m = 0$$

and

$$\sum_{i=1}^{n-2} \sum_{j=i+1}^{n-1} Y_{(i)} Y_{(j)} = \sum_{j=2}^{n-1} Y_{(j)} \sum_{i=1}^{j-1} Y_{(i)}.$$

Hence

$$V_n = \mathcal{E} \sum_{j=1}^{n-1} Y_{(i)}^2 + (n - 1)^2 \mathcal{E} Y_{(n)}^2 - p_n^2$$

$$= t^{2/\alpha} \sum_{j=1}^{n-1} \frac{\gamma(j - 2/\alpha, t)}{\Gamma(j)} + (n - 1)^2 \frac{\gamma(n - 2/\alpha, t)}{\Gamma(n)} - p_n^2$$

$$\sim t^{2/\alpha} \frac{\Gamma(n - 2/\alpha)}{\Gamma(n - 1)} \left(\frac{\alpha}{\alpha - 2} + n - 1 \right) - \frac{t^{2/\alpha}}{(\alpha - 1)^2} \cdot \frac{\Gamma^2(n - 1/\alpha)}{\Gamma^2(n - 1)}.$$

To illustrate these results we assume that $\alpha = 2.5$ and $t = 100$, noting that even when N is as large as 44 $\gamma(N, 100) < 10^{-10}$. The complete gamma approximation is thus "exact" for practical purposes for small n. Table 5.4

Table 5.4

n	ρ_n	$V_n^{1/2}$
2	3.758	14.325
3	6.014	16.458
4	7.818	18.025
5	9.381	19.370
6	10.788	20.592
7	12.083	21.731
8	13.291	22.809
9	14.430	23.839
10	15.513	24.830

lists the numerical values of ρ_n and $V_n^{1/2}$ for $n = 2(1)10$. The skewness of the distribution of the reinsurer's outgo (bounded below by zero) is pronounced for the small n-values.

Verbeek (1966) considered the introduction of a stop-loss reinsurance contract for a portfolio that was already reinsuring all individual sums insured in excess of a maximum retention M_0. Assuming that M_0 had been determined by the principal insurer's demand for a given variance in its aggregate claim outgo, might the introduction of a stop-loss reinsurance with priority $c\mathcal{E}X$ and the consequent increase of M_0 to $M > M_0$ result in lower aggregate reinsurance costs for the principal insurer?

For simplicity we assume that $F_M(x, t)$, the d.f. of the aggregate claims after a maximum retention of M has been set on every individual contract, is approximately normal with ith cumulant

$$\kappa_i = tp_i = t\left\{\int_0^M y^i \, dP(y) + M^i[1 - P(M)]\right\}, \qquad i = 1, 2, 3, \ldots . \quad (5.35)$$

Then the net reinsurance premium for the aggregate claims in excess of $c\mathcal{E}X = ctp_1$ is

$$\int_{ctp_1}^{\infty} (x - ctp_1) \, dF_M(x, t) = (tp_2)^{1/2} \int_{z_0}^{\infty} (z - z_0) \, d\Phi(z) \quad (5.36)$$

and the corresponding sum of squares is

$$\int_{ctp_1}^{\infty} (x - ctp_1)^2 \, dF_M(x, t) = tp_2 \int_{z_0}^{\infty} (z - z_0)^2 \, d\Phi(z), \quad (5.37)$$

where $z_0 = (c - 1)tp_1/(tp_2)^{1/2}$ is a positive quantity. The arguments employed in deriving (3.22) and (3.23) show that

$$\int_{z_0}^{\infty} z \, d\Phi(z) = \left(\int_{-\infty}^{\infty} - \int_{-\infty}^{z_0} \right) z \, d\Phi(z) = -\int_{-\infty}^{z_0} z \, d\Phi(z) = \Phi'(z_0)$$

and

$$\int_{z_0}^{\infty} z^2 \, d\Phi(z) = \left(\int_{-\infty}^{\infty} - \int_{-\infty}^{z_0} \right) z^2 \, d\Phi(z$$

$$= 1 - \int_{-\infty}^{z_0} z^2 \, d\Phi(z) = 1 - \Phi(z_0) + z_0 \, \Phi'(z_0).$$

Hence the net reinsurance premium for aggregate claims in excess of ctp_1 and the variance of the reinsurance payments are, respectively,

$$p_v = (tp_2)^{1/2} \{ \Phi'(z_0) - z_0 [1 - \Phi(z_0)] \} \equiv \sqrt{\kappa_2} \cdot f_1(z_0) \tag{5.38}$$

and

$$\sigma_\rho^2 = tp_2 \{ 1 - \Phi(z_0) + z_0 \, \Phi'(z_0) - 2z_0 \, \Phi'(z_0) + z_0^2 [1 - \Phi(z_0)] \} - p_v^2$$

$$= \kappa_2 [1 - \Phi(z_0) - z_0 f_1(z_0) - f_1^2(z_0)] \equiv \kappa_2 f_2^2(z_0). \tag{5.39}$$

Turning now to the portfolio retained by the principal insurer after effecting the stop-loss reinsurance, we see that the mean outgo is

$$\kappa_1(c) = \int_0^{ctp_1} x \, dF_M(x, t) + ctp_1 \int_{ctp_1}^{\infty} dF_M(x, t)$$

$$= \kappa_1 - \int_{ctp_1}^{\infty} (x - ctp_1) \, dF_M(x, t) = \kappa_1 - p_v \tag{5.40}$$

and the variance is

$$\kappa_2(c) = \int_0^{ctp_1} [x - \kappa_1(c)]^2 \, dF_M(x, t) + [ctp_1 - \kappa_1(c)]^2 \int_{ctp_1}^{\infty} dF_M(x, t)$$

$$= \int_0^{\infty} (x - \kappa_1)^2 \, dF_M(x, t) + p_v^2$$

$$- \int_{ctp_1}^{\infty} \{ [x - \kappa_1(c)]^2 - [ctp_1 - \kappa_1(c)]^2 \} \, dF_M(x, t)$$

$$= \kappa_2 + p_v^2 - \int_{ctp_1}^{\infty} \{ (x - ctp_1)^2 + 2[ctp_1 - \kappa_1(c)](x - ctp_1) \} \, dF_M(x, t)$$

$$= \kappa_2 + p_v^2 - [(\sigma_\rho^2 + p_v^2) + 2(c - 1)\kappa_1 p_v + 2p_v^2]$$

$$= \kappa_2 [1 - f_2^2(z_0) - 2z_0 f_1(z_0) - 2f_1^2(z_0)]$$

$$= \kappa_2 [f_2^2(z_0) + 2 \Phi(z_0) - 1] \equiv \kappa_2 f_3^2(z_0). \tag{5.41}$$

We now assume with Verbeek that the cost of reinsurance is a proportion ε_1 of the net excess-loss premium, *plus* a proportion ε_2 of the standard deviation of the stop-loss reinsurance payments. The cost of the combined reinsurance treaty is then

$$C(M, c) = \varepsilon_1 t \int_M^\infty (y - M) \, dP(y) + \varepsilon_2 \sigma_\rho, \qquad (5.42)$$

and this is to be minimized subject to the invariance of $\kappa_2(c)$, that is, of

$$\kappa_2(c) = \kappa_2 f_3^2(z_0) = \kappa_2 \big|_{M=M_0}. \qquad (5.43)$$

Writing

$$A = C(M, c) + \mu(\kappa_2^{1/2}(c) - \kappa_2^{1/2} \big|_{M=M_0}),$$

where μ is a Lagrangian multiplier, we have

$$\frac{\partial A}{\partial M} = -\varepsilon_1 t[1 - P(M)] + \sqrt{\kappa_2} \, [\varepsilon_2 f_2'(z_0) + \mu f_3'(z_0)] \frac{\partial z_0}{\partial M}$$

$$+ \tfrac{1}{2}\kappa_2'(\kappa_2)^{-1/2}[\varepsilon_2 f_2(z_0) + \mu f_3(z_0)]$$

and

$$\frac{\partial A}{\partial c} = \sqrt{\kappa_2} \, [\varepsilon_2 f_2'(z_0) + \mu f_3'(z_0)] \frac{\partial z_0}{\partial c},$$

where

$$\kappa_2' = \frac{d\kappa_2}{dM} = 2tM[1 - P(M)].$$

Equating these relations to zero and eliminating μ, we have

$$\frac{\varepsilon_1 \sqrt{\kappa_2}}{\varepsilon_2 M} = f_2(z_0) - f_3(z_0) \frac{f_2'(z_0)}{f_3'(z_0)} \qquad (5.44)$$

or, using (5.43) and noting that $f_1'(z_0) = \Phi(z_0) - 1$,

$$\frac{\varepsilon_1 \sqrt{\kappa_2}\big|_{M=M_0}}{\varepsilon_2 M} = \frac{f_3(z_0)}{f_2(z_0)} \left[f_2^2(z_0) - f_3^2(z_0) \frac{2 f_2(z_0) f_2'(z_0)}{2 f_3(z_0) f_3'(z_0)} \right]$$

$$= \frac{f_3(z_0)}{f_2(z_0)} \left\{ f_2^2(z_0) + f_3^2(z_0) \frac{f_1(z_0) \, \Phi(z_0)}{[1 - \Phi(z_0)][f_1(z_0) + z]} \right\}$$

$$= \frac{f_3(z_0)}{f_2(z_0)} \left\{ 1 - 2\,\Phi(z_0) + f_3^2(z_0) \frac{\Phi'(z_0)}{[1 - \Phi(z_0)][z_0\,\Phi(z_0) + \Phi'(z_0)]} \right\}.$$

$$(5.45)$$

Verbeek (loc. cit.) illustrates the application of relations (5.45) and (5.43), where $z_0 = (c - 1)tp_1/(tp_2)^{1/2}$, to the case

$$P'(y) = \tfrac{1}{2}\alpha^3 y^2 e^{-\alpha y}, \qquad 0 < y < \infty,$$

with $\alpha = 6 \times 10^{-5}$ (so that $p_1 = 10^5/2$, $p_2 = 10^{10}/3$), $\varepsilon_1 = 0.05$, $\varepsilon_2 = 0.5$, and $M_0 = 115 \times 10^3$. A brief summary of his results is given in Table 5.5. It will be noticed that the minimum cost of reinsurance occurs for M approximately 300,000 in a period when expected claims amount to 400. When the

Table 5.5

	$t = 400$		$t = 795$	
$10^{-3}M$	c	$10^{-3}C$	c	$10^{-3}C$
115	∞	342.5	∞	681.0
200	1.021	333.5	1.015	537.0
300	1.006	326.5	1.004	476.5
400	1.002	330.5	1.002	469.0
500	1.001	333.0	1.001	470.5
∞	1.000	337.0	1.000	475.0

"period" lengthens to 795 claims, the (higher) cost is a minimum near $M = 400,000$. A disturbing feature of this illustration is that with $\sqrt{\kappa_2}/\kappa_1 = 2\sqrt{t/3}$ equal to 23.1 and 32.6, respectively, c is so close to unity. This implies that all outgo in excess of the mean $\mathcal{E}X$ is subject to the stop-loss reinsurance.

REFERENCES

Ammeter, H. (1946). "Das Maximum des Selbstbehaltes in der Lebensversicherung unter Berücksichtigung der Rückversicherungen." *Mitt. Verein. Schweiz. Versich. Mathr.*, **46**, 187–213.

——— (1954). "Risikotheoretische Methoden in der Rückversicherung." *Mem. XIV Cong. Intern. Actu.*, Madrid, **1**, 612–632.

——— (1964). "The rating of 'largest claim' reinsurance covers." *Quarterly Letter*, Jubilee No. 2, 5–17. Algemeene Reins. Co., Amsterdam.

Benktander, G. (1964). "New forms of excess of loss reinsurance." *Trans. XVII Intern. Cong. Actu.*, London, **3**, 502–514.

Benktander G., and J. Ohlin (1967). "A combination of surplus and excess reinsurance of a fire portfolio." *Astin Bull.*, **4**, 177–190.

Berger, A. (1925). *Die Prinzipien der Lebensversicherungstechnik*. Vol. II. Springer, Berlin.

Berger, G. (1964). "Zur praktischen Durchführung Risikotheoretischer Untersuchungen." *Trans. XVII Intern. Cong. Actu.*, London, **3**, 516–524.

Cacciafesta, R. (1962). "Sulla riassicurazione per eccesso sinistro singolo." *Gior. Ist. Ital. Attu.*, **25**, 187–198.

Deitz, R. (1954). "Problèmes d'actuariat que pose la réassurance sur la vie, spécialement détermination des pleins de conservation." *Mem. XIV Cong. Intern. Actu.*, Madrid, **1**, 451–468.

Dubordieu, J. (1952). *Théorie Mathématique du Risque dans les Assurances de Répartition*. Gauthier-Villars, Paris.

Finetti, B. De (1940). "Il problema dei 'pieni'." *Gior. Ist. Ital. Attu.*, **11**, 1–88.

Hultmann, K. (1942). "Einige numerische Untersuchungen auf Grund der kollektiven Risikotheorie." *Skand. Aktu. Tidskr.*, **25**, 84–119, 169–199.

Lambert, H. (1960). "Une application de la théorie collective du risque: la réassurance." *Bull. Assoc. R. Actu. Belges*, **60**, 33–47.

Lundberg, O. (1954). "Actuarial problems appertaining to reinsurance of disability risks." *Mem. XIV Cong. Intern. Actu.*, Madrid, **1**, 844–856.

Magnus, W., F. Oberhettinger, and R. P. Soni (1966). *Formulas and Theorems for the Special Functions of Mathematical Physics*. Springer, Berlin.

Medolaghi, P. (1909). "La teoría del rischio e le sue applicazioni." *Ber. VI Intern. Kong. Versich. Wissens.*, Vienna, **1**, 723–746.

Meidell, B. (1953). "Randbemerkungen zum Landréschen Maximum." *Skand. Aktu. Tidskr.*, **36**, 168–181.

Ottaviani, G. (1952). "Sul problema della riassicurazione." *Gior. Ist. Ital. Attu.*, **15**, 65–84.

Pentikäinen, T. (1952). "On the net retention and solvency of insurance companies." *Skand. Aktu. Tidskr.*, **35**, 71–92.

——— (1954). "On the reinsurance of an insurance company." *Mem. XIV Cong. Intern. Actu.*, Madrid, **1**, 389–400.

Richard, P. (1949). "Sur un nouveau mode de réassurance." *Bull. Trim. Inst. Actu. Franç.*, **48**, 344–367.

Segerdahl, C.-O. (1948). "Some properties of the ruin function in the collective theory of risk." *Skand. Aktu. Tidskr.*, **31**, 46–87.

Shirasugi, S. (1954). "Die Theorie und Praxis der Lebensrückversicherung." *Mem. XIV Cong. Intern. Actu.*, Madrid, **1**, 433–442.

Thépaut, A. (1950). "Le traité d'excédent du coût moyen relatif (ECOMOR)." *Bull. Trim. Inst. Actu. Franç.*, **49**, 273–343.

——— (1953). "Essai de détermination pratique du plein de conservation." *Bull. Trim. Inst. Actu. Franç.*, **52**, 371–472.

——— (1957). "Calcul de plein de conservation dans l'assurance des corps de navire." *Trans. XV Cong. Intern. Actu.*, New York, **2**, 398–406.

Verbeek, H. G. (1966). "On optimal reinsurance." *Astin Bull.*, **4**, 29–38.

Wagner, K. (1898). *Das Problem vom Risiko in der Lebensversicherung*. Fischer, Jena.

CHAPTER 6

Utility Theory and Its Application to Reinsurance and Profit Prospects

LOGARITHMIC UTILITY OF MONEY

Only a miser can take as much pleasure in adding a dollar to a fortune of $10 million as he had when he increased his assets of $1000 by $1. Daniel Bernoulli (1738) accordingly assumed that in ordinary circumstances the relative value of a capital increment varies directly as the increment and inversely as the existing capital. If u is a *fortune morale*, in Laplace's words, and x, the corresponding *fortune physique*, Bernoulli's assumption is that

$$du = k\frac{dx}{x}, \qquad x > 0,$$

or

$$u = k \ln x + h, \tag{6.1}$$

where h is the *fortune morale* corresponding to a unit of actual money. Today Laplace's *fortune morale* is called "utility" and u would be called the utility of a sum of money x. An interesting historical discussion of the concept of utility is found in Savage (1954, Chapter 5).

[The mathematics of Bernoulli's article is provided by Todhunter (1865) and an embellished account is found in Chapter 10 of Laplace (1812). DeMorgan's (1845) encyclopaedia article has been referred to as essentially a translation of Laplace's text and his paragraphs 163 *et seq*. refer to the relative value (utility) of money. Very few modern probability texts make more than passing reference to Bernoulli's "moral expectation."]

Suppose that a person with a capital of a has a probability p_j of increasing it by b_j $(j = 1, 2, \ldots, n)$, the n increases being mutually exclusive. The person's moral expectation is then

$$u = k\sum_{j=1}^{n} p_j \ln (a + b_j) + h = k \ln x + h,$$

where x is the corresponding monetary value. Hence

$$x = \prod_{j=1}^{n} (a + b_j)^{p_j}. \tag{6.2}$$

As an application of (6.2), consider the insurance of some goods valued at s and expected to arrive from abroad by ship. Let $p = 1 - q$ be the probability of the goods' safe arrival and a, the value of the policyholder's capital. Can the insured afford to pay more than the net premium sq for the insurance of his goods? We shall show that the answer is, Yes.

The result of insuring the goods on a net premium basis is that the policyholder has a guaranteed sum of $a + s - sq = a + sp$. If the business-man did not insure the goods, his *fortune physique* by (6.2) would be $(a + s)^p a^q$, and this is the geometric mean of the quantities $a + s$ and a with "weights" p and q, respectively. This geometric mean is less than the corresponding arithmetic mean, namely $p(a + s) + qa = a + sp$, and there-fore the businessman can afford to pay more than sq for the insurance. The precise amount he may pay without disadvantage depends on his capital a. Thus, if the insurance company imposes a loading at rate η on the net premium sq, the businessman would find it indifferent whether he insured the goods or not if his capital a were determined from the equation

$$(a + s)^p a^q = a + s - sq(1 + \eta). \tag{6.3}$$

Bernoulli's own illustration was to assume that $s = 10^4$, $p = 0.95$, and $\eta = 0.6$—a rather high risk and expense loading! Equation 6.3 then becomes

$$(10^4 + a)^{0.95} a^{0.05} = a + 10^4 - 800,$$

and this is satisfied approximately by $a = 5043$, which indicates that if the businessman's capital exceeds 5043 he should not insure his goods.

If, on the other hand, we inquire what risk reserve R the insurance company should possess to offer to insure a simple risk of the type described, this is given as the solution of the equation

$$(R + sq\overline{1 + \eta})^p (R + sq\overline{1 + \eta} - s)^q = R. \tag{6.4}$$

Using the same illustrative values for s, p, and η, we find, with Bernoulli, that $R = 14,243$.

Bernoulli also enunciated a proposition that we may paraphrase as follows. It is more advantageous to accept n independent risks of losing s/n each with probability q than a single risk of losing s with probability q. Laplace was able to prove this in his Chapter 10.

The foregoing ideas were utilized by Barrois (1835) in the third part of his mathematical treatise on fire insurance. His principal contributions were the generalization of (6.3) to (a) the insurance of three contiguous buildings of unequal value, given the probabilities p_i $(i = 1, 2, 3)$ of their destruction by fire during the year, and the six probabilities p_{ij} $(i \neq j = 1, 2, 3)$ of a fire in building i spreading to building j, and (b) the insurance of a long warehouse, given the probability of an outbreak of fire at any point and assuming the d.f. of the subsequent length which would be destroyed to be $1 - e^{-ay}$.

Mounier (1894) expanded the right-hand side of (6.2) in powers of b_j/a and ignored third and higher powers, thus obtaining

$$x = \sum_{j=1}^{n} p_j b_j - \frac{1}{2a} \left[\sum_{j=1}^{n} p_j b_j^2 - \left(\sum_{j=1}^{n} p_j b_j \right)^2 \right]$$

namely, the expected increase in the aleatory capital, minus a fraction of its variance. He applied this relation to an individual seeking life insurance coverage and concluded that the contract would be advantageous only if the net premium loading were less than a given fraction of the contract's variance.

THE UTILITY INDEX OF A RISK BUSINESS

We now generalize Bernoulli's logarithmic utility to a utility function of money $u(x)$, which must be nondecreasing for increasing x and presumably concave to the X-axis at least for large values of x. We allow $u(x)$ to exist for negative sums of money x and assume that any risk business has some specified utility function $u(x)$ on which it can base its decisions about risk taking and profit preferences. We define the utility of a risk situation characterized by the pair $\{R, F(\cdot)\}$—R being the company's risk reserve and $F(\cdot)$, its aggregate claim d.f.—by means of the utility index $U(R, F(\cdot))$ defined by

$$U(R, F(\cdot)) = \int_0^\infty u(R + \pi_1 - x) \, dF(x), \qquad (6.5)$$

where π_1 is the loaded aggregate premium for the period considered. We may regard (6.5) as a natural definition of the utility of a risk situation, namely, as the expected value of the utility of the various risk outcomes, or it can be obtained as the result of a series of axioms (Borch, 1960b; Wolff, 1966). Note that when

$$u(x) = k \ln x + h \quad \text{and} \quad F(x) = \begin{cases} p, & x = 0, \\ 1, & x = s, \end{cases}$$

relation (6.4) follows from (6.5) if $U(R, (F \cdot)) = u(R)$, namely, if we are prepared to sell the business for its capital (risk reserve) value.

The logarithmic utility assumption does not adapt easily to the general case of (6.5). As an alternative exponential utility (Borch, 1963; Ferra, 1964; Welten, 1964; Wolff, 1966) assumes

$$u(x) = a - be^{sx} \qquad (6.6)$$

and thus

$$U(R, F) = \int_0^\infty [a - be^{s(R+\pi_1-x)}] \, dF(x)$$
$$= a - be^{s(R+\pi_1)} \, \theta(s), \tag{6.7}$$

where $\theta(\cdot)$ is the L.S.T. of $F(\cdot)$. We may ascribe numerical values to the parameters a, b, and s by formulating three boundary conditions on $u(\cdot)$; for example (Welten, 1964), we could write

$$u(R + \pi_1 - \kappa_1) = u(R + \eta_1) = 1 \quad \text{and} \quad u(R) = 0,$$

meaning that a year's operations at the "expected" level is our unit of utility, and a year's "standstill" is of no utility. Ruin of the business might be given a negative utility; for example,

$$u(0) = -\frac{R}{\eta_1} - k,$$

where k might be 30 or 40 and $u(0)$ would then represent the "disutility of ruin."

However, most published work on the application of utility theory to a risk business has been based on Borch's (1960b) quadratic assumption, namely,

$$u(x) = ax(b - x), \qquad a > 0, x \le \frac{b}{2}, \tag{6.8}$$

in which the utility of no money is zero and the restriction to values of $x \le b/2$ is to prevent $u(x)$ decreasing. Relation 6.5 then becomes

$$a^{-1}U(R, F) = \int_0^\infty (R + \pi_1 - x)(b - \overline{R + \pi_1} + x) \, dF(x)$$
$$= \int_0^\infty (R + \eta_1 - \overline{x - \kappa_1})(b - \overline{R + \eta_1} + \overline{x - \kappa_1}) \, dF(x)$$
$$= (R + \eta_1)(b - \overline{R + \eta_1}) - \kappa_2, \tag{6.9}$$

that is,

$$U(R, F) = u(R + \eta_1) - a\kappa_2, \tag{6.10}$$

where $R + \eta_1$ is the expected risk reserve after the prospective period's operations and κ_2 is the variance of X. If we require that the utility index be nondecreasing, its use must be limited to situations in which $R + \eta_1 \le b/2$. The maximum utility index is then $a(b^2/4 - \kappa_2)$.

We notice from (6.10) that the utility of a risk business is less than the utility of a guaranteed sum of money equal to the expected year-end risk reserve, the difference being a times the variance of the aggregate outgo of the portfolio. The quadratic utility assumption is thus equivalent to a demand that a risk business seek to minimize the variance of its outgo, having received a risk premium loaded by an amount η_1.

THE UTILITY OF REINSURANCE

Let us now take the case in which a reinsurer agrees to accept any portion of a principal insurer's portfolio subject to a loading of $100\,\varepsilon\%$ of the corresponding net risk premium. Consider first the problem of finding an optimal uniform quota α of every contract's outgo to be reinsured. If the utility of money is given by (6.8), the risk business's utility index after reinsurance is

$$\int_0^\infty u(R + \pi_1 - \overline{1 + \varepsilon}\alpha\kappa_1 - \overline{1 - \alpha}x)\,dF(x)$$

$$= a\int_0^\infty [R + \eta_1 - \alpha\varepsilon\kappa_1 - (1 - \alpha)(x - \kappa_1)]$$

$$\times [b - R - \eta_1 + \alpha\varepsilon\kappa_1 + (1 - \alpha)(x - \kappa_1)]\,dF(x)$$

$$= a(R + \eta_1 - \alpha\varepsilon\kappa_1)(b - \overline{R + \eta_1 - \alpha\varepsilon\kappa_1}) - a(1 - \alpha)^2\kappa_2, \quad (6.11)$$

where

$$\kappa_1 = \int_0^\infty x\,dF(x) \quad \text{and} \quad \kappa_2 = \int_0^\infty (x - \kappa_1)^2\,dF(x).$$

Differentiating (6.11) with respect to α, we obtain

$$2a\left[\left(R + \eta_1 - \frac{b}{2}\right)\varepsilon\kappa_1 + \kappa_2 - (\kappa_2 + \varepsilon^2\kappa_1^2)\alpha\right],$$

which has a negative derivative. Equating this result to zero, we have the condition for a maximum, namely

$$\alpha = (\kappa_2 + \varepsilon^2\kappa_1^2)^{-1}\left[\kappa_2 - \left(\frac{b}{2} - \overline{R + \eta_1}\right)\varepsilon\kappa_1\right]. \quad (6.12)$$

When $\varepsilon > \kappa_2/(b/2 - \overline{R + \eta_1})\kappa_1$, quota reinsurance would not improve the insurance company's utility index.

Borch (1961b) provides a numerical example in which $\kappa_1 = \kappa_2 = 1$, $b = 3$, $a = \frac{1}{3}$, $R = 0$, $\eta_1 = 0.2$, and $\varepsilon = 0.1$, and calculates the utility index, $U(R, F)$, the expected profit, $\eta_1 - \alpha\varepsilon\kappa_1$, and the probability of ruin at the end of the year, viz. $1 - F(\overline{1 - \alpha}^{-1}(R + \pi_1 - \overline{1 + \varepsilon}\alpha\kappa_1))$, for the special case $F(x) = 1 - e^{-x}$, with $\alpha = 0.0(0.1)1.0$.

On the other hand, suppose that the risk business decides to effect a stop-loss reinsurance of all aggregate outgo $X > c$. The net stop-loss reinsurance premium would then be

$$\rho = \int_c^\infty (x - c)\, dF(x) \tag{6.13}$$

and the risk business's utility index after effecting the reinsurance treaty would be

$$\int_0^c u\big(R + \pi_1 - \overline{1 + \varepsilon\rho} - \min(c, x)\big)\, dF(x)$$

$$= \int_0^c u(R + \pi_1 - \overline{1 + \varepsilon\rho} - x)\, dF(x)$$

$$+ u(R + \pi_1 - \overline{1 + \varepsilon\rho} - c)[1 - F(c)]. \tag{6.14}$$

When we use the quadratic assumption (6.8), the first (integral) term of (6.14) becomes

$$a\int_0^\infty (R + \eta_1 - \overline{1 + \varepsilon\rho} - \overline{x - \kappa_1})(b - \overline{R + \eta_1} + \overline{1 + \varepsilon\rho} + \overline{x - \kappa_1})\, dF(x)$$

$$- a\int_c^\infty (R + \pi_1 - c - \overline{1 + \varepsilon\rho} - \overline{x - c})$$

$$\times (b - \overline{R + \pi_1} + c + \overline{1 + \varepsilon\rho} + \overline{x - c})\, dF(x)$$

$$= a(R + \eta_1 - \overline{1 + \varepsilon\rho})(b - \overline{R + \eta_1} + \overline{1 + \varepsilon\rho})$$

$$- a\kappa_2 - u(\overline{R + \pi_1} - \overline{1 + \varepsilon\rho} - c)[1 - F(c)]$$

$$- a[2(\overline{R + \pi_1} - c - \overline{1 + \varepsilon\rho}) - b]\rho + a\int_c^\infty (x - c)^2\, dF(x).$$

The utility index (6.14) is thus equal to

$$a(R + \eta_1 - \overline{1 + \varepsilon\rho})(b - \overline{R + \eta_1} + \overline{1 + \varepsilon\rho}) - a\kappa_2$$

$$- a[2(R + \pi_1 - c - \overline{1 + \varepsilon\rho}) - b]\rho + a\int_c^\infty (x - c)^2\, dF(x). \tag{6.15}$$

We notice that this differs from the direct application of (6.10), since the last three terms are not equal to minus a times the variance of the retained portfolio. We now require the maximum of (6.15) for varying c. Differentiating with respect to c, remembering that ρ is a function of c, we obtain

$$a[1 - F(c)]\{(1 + \varepsilon)[b - 2(R + \eta_1) + 2\rho(1 + \varepsilon)]$$

$$+ 2[R + \pi_1 - c - 2(1 + \varepsilon)\rho] - b\},$$

and this is equal to zero either when $F(c) = 1$, that is, no reinsurance effected, or when

$$\varepsilon[b - 2(R + \eta_1)] + 2\kappa_1 = 2(1 + c - \varepsilon^2\rho). \tag{6.16}$$

Consideration of the second derivative of (6.15) shows that no reinsurance corresponds to a minimum, whereas (6.16) is the condition for a maximum. Since the right-hand side of (6.16) is always increasing as c increases, there is only one value of c that satisfies the condition for a maximum utility index.

We note that in both examples the reinsurer's loading rate ε is unrelated to η_1, the cedent's over-all loading. The cost of the reinsurance can thus exceed the margin in the corresponding portion of the cedent's gross premium, and the desirability of reinsurance in such a case is automatically allowed for by the assumed utility function.

Reinsurance of any type without cost to the cedent other than the amount of the risk premium ρ (i.e., $\varepsilon = 0$) is equivalent to the principal insurer refusing to accept certain risks. That is to say, the risk business is prepared to forego a given portion of its expected profit η_1 by refusing to accept risks corresponding to a fixed amount of risk premium equal to ρ. The problem of selecting the "best" type of "reinsurance" under these circumstances was first considered by Borch (1960a), who proved that a stop-loss contract with ρ given by (6.13) would result in a minimum variance for the retained portfolio. An elegant proof of a generalized statement of this theorem was provided by Kahn (1961). Further generalizations have been published by Pesonen (1967) and Ohlin (1968), the latter proving that the risk business can gain nothing in the way of variance reduction by allowing its reinsurance policy to depend on the individual claims.

We mention that Vajda (1963) showed that the reinsurance company's variance is minimized for given fixed reinsurance risk premium ρ by a quota reinsurance, the quota α being determined from $\rho = \alpha \kappa_1$. Hickman and Zahn (1966) provide an alternative proof. When the reinsurance company has a given (approximate) probability of ultimate ruin (calculated according to (4.64) with the first factor set equal to unity), Lambert (1963) has shown that the loading rates η_r on its risk premiums must be in increasing order of size when its portfolio consists entirely of proportionate, excess-loss, or stop-loss reinsurance, respectively, accepted from a risk business with given $F(\cdot)$. This order of preference is exactly the opposite of the order preferred by the cedent and emphasizes that the interests of cedent and reinsurer are opposed when each tries to decrease his probability of ruin.

PROFIT PROSPECTS AND DIVIDENDS TO STOCKHOLDERS

In Chapter 4 no thought was given to the possibility of paying dividends to stockholders out of the risk reserve. It was merely understood that interest earnings on the reserve could be added to the product of any profit loadings in the premium that were specifically intended for distribution to the stockholders. If $R(\tau)$ became "too large" at any moment τ, as measured by the resulting probability of ruin within a shorter or longer period thereafter, the theory of Chapter 4 would have met the situation by reducing the loading η, whereas that of Chapter 5 would have increased the maximum retention M or the priority c of a stop-loss reinsurance treaty. However, if we wish to think in terms of profits to stockholders over and above a specific profit loading in every premium received, we could agree to "skim the cream" from the risk reserve whenever it exceeded the amount necessary to keep the probability of ruin within t years below a specified small value.

We saw in Chapter 4 that if we fix a specified ceiling Z on the risk reserve so that, once it is attained, we conserve only so much of the year's excess income over outgo as is necessary to "top up" the reserve to Z, the eventual ruin of the risk business becomes a certainty. Nevertheless, we may impose such a ceiling Z and calculate the expected time to ruin or we may consider the value of the expected dividends paid from the risk reserve after it reaches Z and before ruin occurs (Finetti, 1957).

Let $D(R)$ be the expected number of years before the ruin of a risk business with a d.f. of aggregate claim outgo equal to $F(\cdot)$ and, as usual, write $\pi_1 = \kappa_1 + \eta_1$, where $\kappa_1 = \int_0^\infty x \, dF(x)$. Then, $0 \le R < Z$,

$$D(R) = 1 + \int_0^{R+\pi_1} D(R + \pi_1 - x) \, dF(x), \qquad (6.17)$$

with the boundary condition $D(R) = D(Z)$, $R \ge Z$. This is an integral equation that can be solved numerically for given $F(\cdot)$ (Kopal, 1961). Borch (1965) has provided an explicit general solution for the case $F(x) = 1 - e^{-x}$ by the use of Laplace transforms. Finetti's (1957) original example was the random walk expressed by $\pi_1 = 1$ and

$$F(x) = \begin{cases} p, & 0 \le x < 2, \\ 1, & 2 \le x < \infty. \end{cases} \qquad (6.18)$$

Equation 6.17 then becomes

$$D(R) = 1 + p \, D(R + 1) + q \, D(R - 1), \qquad (6.19)$$

with the boundary conditions

$$D(0) = 0 \quad \text{and} \quad D(Z) = 1 + p \, D(Z) + q \, D(Z - 1).$$

For $p > q$ the solution of (6.19) is (cp. Feller, 1968, cited in Chapter 3)

$$D(R) = \frac{p}{(p - q)^2} \left(\frac{p}{q}\right)^Z \left[1 - \left(\frac{q}{p}\right)^R\right] - \frac{R}{p - q} . \qquad (6.20)$$

A numerical illustration in Borch (1964) shows how rapidly $D(\cdot)$ increases with Z.

Turning now to the evaluation of future dividends, we note that the dividend at time τ ($\tau = 0, 1, 2, \ldots$) is $d_\tau = \max\left(0, R(\tau) - Z\right)$ and thus a random variable. Let us discount future dividends at an interest rate i with $v = (1 + i)^{-1}$ and write $R(0) \equiv R$.

We will suppose that the interest earned on the risk reserve is at a rate such that the value of a unit payable at the end of a year is equal to $1 - k(1 - v)$, $0 \le k \le 1$. This interest is always assumed paid to the stockholders, whatever the size of R; it is further supposed that the net profit of any year, namely $\pi_1 - X$, where X is the aggregate claim outgo with d.f. $F(\cdot)$, and $\pi_1 = \kappa_1 + \eta_1$, where $\kappa_1 = \int_0^\infty x \, dF(x)$, does not earn any interest. Under these circumstances the following recursive relation holds for $V(R)$, the value of future dividends, $0 < R \le Z$:

$$V(R) = k(1 - v)R + v \int_0^{R+\pi_1-Z} [V(Z) + R + \pi_1 - Z - x] \, dF(x)$$

$$+ v \int_{R+\pi_1-Z}^{R+\pi_1} V(R + \pi_1 - x) \, dF(x), \quad (6.21)$$

where the first integral, which represents the dividend payment made when X is small, vanishes when $R \le Z - \pi_1$, and in such case the lower limit of the second integral is zero. This equation may be simplified to

$$V(R) = k(1 - v)R + v \, V(Z) \, F(R + \pi_1 - z)$$

$$+ v \int_0^{R+\pi_1-Z} F(x) \, dx + v \int_0^Z V(x) \, d_x F(R + \pi_1 - x), \quad (6.22)$$

with the second and third terms zero when $R \le Z - \pi_1$ and the upper limit Z of the final integral then replaced by $R + \pi_1$. When $R > Z$, the boundary equation

$$V(R) = V(Z) + R - Z \quad (6.23)$$

applies. Only very special cases of (6.22) have been treated in the literature to date.

Borch (1965) derived a similar integral equation* for the case in which $k = 0$ and X is not restricted to the positive real axis. Using Laplace transforms, he obtained an explicit result for the special case $F(x) = 1 - e^{-x}$, $0 < x < \infty$. In a later (1966a) article he considered the density

$$F'(x) = \begin{cases} abe^{-b(\pi_1-x)}, & x < \pi_1 \\ (1 - a)be^{b(\pi_1-x)}, & x > \pi_1 \end{cases} \qquad \tfrac{1}{2} < a < 1,$$

and illustrated his explicit solution with the values $a = 0.603$, $b = 1$, and $v = 0.97$, by calculating $V(R)$ for $R, Z = 0(1)5$.

---

* Borch's earlier treatment (not cited here) appears to apply only when R is approximately equal to Z.

In the Finetti (1957) case, in which $\pi_1 = 1$, $F(\cdot)$ is given by (6.18) and R is integral, (6.22) simplifies to

$$V(R) = k(1 - v)R + vp\,V(R + 1) + vq\,V(R - 1) \qquad (6.24)$$

for $0 < R \leq Z$, whereas (6.23) provides

$$V(Z + 1) = V(Z) + 1$$

and at the other boundary $V(0) = 0$. Standard methods for solving linear difference equations (e.g., Feller, *loc. cit.*) provide the solution of (6.24), namely,

$$V(R) = k\left(R + \frac{p - q}{i}\right) + A\alpha^R - \left(A + \frac{k}{i}\,\overline{p - q}\right)\beta^R, \qquad (6.25)$$

where α and β are the roots of the quadratic equation

$$pz^2 - (1 + i)z + q = 0, \qquad \alpha > 1, \quad \beta < 1,$$

and

$$A = \frac{1 - k - (k/i)(p - q)(1 - \beta)\beta^Z}{(\alpha - 1)\alpha^Z + (1 - \beta)\beta^Z}.$$

Viewed as a function of Z, the right-hand-side of (6.25) depends on the behavior of A. Finetti (1957) has studied this in detail for the three cases (a) $k = 0$ (as in Borch's papers), (b) $k = 1$ (interest earned on the risk reserve, the same as that used in discounting dividends), and (c) $0 < k < 1$. In the last case the risk business maintains a risk loading of $\eta_1 > 0$ only as long as $k \geq 1 - \dfrac{p - q}{i}$.

Borch (1964) reconsidered (6.24) for $k = 0$ with the slightly different boundary condition $V(-1) = 0$, that is, ruin at $R = -1$ instead of at $R = 0$. He proved that it was not possible to increase $V(R)$ by proportionate (quota) reinsurance and illustrated this numerically by calculating $V(R)$ for $R = 0(1)5$ and $Z = 0(1)6$ with $q = 0.435$ and $v = 0.983$. Because quota reinsurance leads to an increase in the $D(R)$ of (6.19) (leading to an infinite life expectancy—and no dividends—when the reinsured quota is 100%) Borch suggests that insurance company management should seek to maximize some function of both $V(R)$ and $D(R)$ when it formulates a reinsurance policy. In (1966b) this author extended (6.25) with $k = 0$ to include the case in which R need not be an integer at a given point of time.

RECIPROCAL REINSURANCE TREATIES

We have seen that a risk business can decrease its probability of ruin by reinsuring a certain portion of its portfolio of risks. If the cost of this reinsurance is too high, the company's utility index may be decreased as a result of the transaction and its natural reaction would be to seek reinsurance elsewhere. The question then arises whether a reinsurance treaty can be profitable in some sense for both cedent and reinsurer. A related question is whether there are mutual advantages in reciprocal reinsurance between two or more companies.

This type of question was considered by Finetti (1942), who was the first to apply utility (or, as he called them, preference) functions to the partial interchange of insurance portfolios. Suppose n companies desire to improve their utility indices by mutual reinsurance of part of each other's portfolios. Write s_i $(i = 1. 2, \ldots, n)$ for some measure of the portfolio of company i and σ_i^2 for the variance of the portfolio's outgo, assumed to be proportionate to its size. Then Finetti assumed company i's utility index (preference function) to be of the form (6.9) without particular reference to its risk reserve R_i. He wrote, in effect,

$$U_i(s_i) = s_i - \beta_i \sigma_i^2, \qquad \beta_i > 0. \tag{6.26}$$

Suppose that company i cedes to company j an amount s_{ij} of its portfolio so that $s_{ij} \geq 0$ and $s_i = \sum_{j=1}^n s_{ij}$. Its final portfolio is then $t_i = \sum_{k=1}^n s_{ki}$ and has a variance $\bar\sigma_i^2 = \sum_{k=1}^n s_{ki}^2 \sigma_k^2 / s_k^2$. Hence the utility index of the company's portfolio after the reinsurance exchanges have taken place is

$$U_i(t_i) = t_i - \beta_i \bar\sigma_i^2 = \sum_{k=1}^n \left(s_{ki} - \frac{\beta_i s_{ki}^2 \sigma_k^2}{s_k^2} \right)$$

$$= k_i - \beta_i \sum_{k=1}^n \frac{\sigma_k^2}{s_k^2} (s_{ki} - \xi_{ki})^2, \tag{6.27}$$

where $\xi_{ki} = s_k^2 / 2\beta_i \sigma_k^2$ and $k_i = t_i + (4\beta_i)^{-1} \sum_{k=1}^n s_k^2 / \sigma_k^2$. Finetti proves that the optimum set of values s_{ki} is given by

$$s_{ki} = \xi_{ki} + \lambda_i \left(s_k - \sum_{j=1}^n \xi_{kj} \right), \tag{6.28}$$

subject to $\lambda_i \geq 0$ and $\sum_{i=1}^n \lambda_i = 1$.

In the particular case in which s_k / σ_k^2 is the same for every company $\xi_{ki} = \alpha_i s_k$ and (6.28) gives

$$s_{ki} = \left[\alpha + \lambda_i \left(1 - \sum_{j=1}^n \alpha_j \right) \right] s_k = \alpha_i s_k = \left(1 - \sum_{j \neq k} \alpha_j \right) s_k,$$

since $\sum_{i=1}^n s_{ki} = s_k$ and thus $\sum_{j=1}^n \alpha_j = 1$. Hence in this special case company k gives company i a proportion of its portfolio s_k that does not depend on i; that is to say, unless $n = 2$, every company gives one nth of its portfolio to each other company and retains one nth itself ("equipartition"). If $n = 2$, quota reinsurance is optimum, with $\alpha_1 + \alpha_2 = 1$.

Finetti (*loc. cit.*) pointed out that excess-loss reinsurance between n companies is nothing more than imposing a quota that varies from policy to policy and would thus also lead to equipartition of the pool of risks between the companies. This problem was taken up again by Borch in a series of six overlapping papers, two of which (1960b, 1961a) related to the two company exchange, and the other four (1960c, 1962a,b,c) to the n company situation. Wolff (1966) has devoted most of his Abschnitt IV to a careful and original mathematical presentation of this work and a lacuna in one of its theorems has been filled by DuMouchel (1968).

As already mentioned, Borch's approach is through a utility index written in the form (6.5) and sometimes specialized to the form (6.9). The most explicit results are obtained (1960b) when each of two companies has the same utility function in the form (6.9). In this case it is shown that mutual quota reinsurance with $\alpha_1 + \alpha_2 = 1$ improves the utility indices of both companies, and this optimum is studied as a function of α_1 and the difference (supposed independent of α_1) between the two companies' reinsurance charges. In the case of more general utility functions $u(x)$, with $u'(x)\searrow$, it is heuristically shown (1961a) that optimality is achieved when and only when

$$u_2'(R_2 - x_1 - x_2 + z) = k u_1'(R_1 - z), \qquad k > 0, \qquad (6.29)$$

where R_1 and R_2 are the respective risk reserves, including the premiums paid for the year, x_1 and x_2 are the aggregate claim outgoes before reinsurance, and z is the amount of the outgo of the first company after exchange of reinsurance has taken place. We note that (6.29) implies that an increase in the first company's utility necessarily leads to a decrease in that of the other. This reveals the conflict of interest that exists between companies exchanging part of their portfolios.

A theorem proved by Wolff (1966) for the two-company case (and extended by him to the n company situation) is that the optimality of quota reinsurance holds only when the utility functions of the companies are related by means of the equation

$$u_2(x) = a + b u_1\left(\frac{1 - \alpha_1}{\alpha_1} x + c\right), \qquad b > 0. \qquad (6.30)$$

There are, of course, utility function pairs that would make it advantageous for one company to transfer the whole of its portfolio to the other.

In the n-company case quota reinsurance results as the optimum (Borch, 1960c) from the quadratic assumption

$$u_i(x) = x - a_i x^2. \qquad (6.31)$$

Borch's papers and Wolff's book should be consulted for further details.

The problem of the conflict of interest between cedent and reinsurer has been considered by Lambert (1963) from the (ultimate) ruin viewpoint without the use of utility functions. As a measure of its "efficiency," each protagonist is supposed to use a function Δ defined as the margin of loading left for profits or dividends to policyholders after subtracting the risk loading required on the retained net premium for the given ruin probability and also, in the case of the cedent, the reinsurer's loading. Writing Δ for the cedent as Δ_c and for the reinsurer as Δ_r, we have

$$\Delta_c = \eta - \eta_c p_1^{(c)} - \eta_r p_1^{(r)}, \qquad p_1 = p_1^{(c)} + p_1^{(r)}, \qquad (6.32)$$

where η is the loading rate applied to p_1, the mean sum-at-risk (namely, unity), and the amount required to achieve the desired probability of ultimate ruin, η_c is the loading rate required by the cedent on the retained portfolio with mean sum-at-risk $p_1^{(c)}$, and η_r is the reinsurer's loading rate applied to a mean sum-at-risk $p_1^{(r)}$.

The probability of ultimate ruin, given a value like 10^{-3}, is to be represented by the approximate asymptotic formula derived from (4.64), namely,

$$\psi(R) \sim e^{-\kappa R},$$

where R is the company's risk reserve. The value of κ thus obtained is then used to determine η from the relation corresponding to (4.62), namely

$$1 + \kappa(1 + \eta) = \int_0^\infty e^{\kappa y} \, dP(y), \qquad (6.33)$$

whereas η_c is found from

$$1 + \kappa(1 + \eta_c)p_1^{(c)} = \int_0^\infty e^{\kappa Ty} \, dP(y), \qquad (6.34)$$

where Ty represents the set of y-values retained by the cedent on its own account; for example, excess-loss reinsurance would be written as

$$Ty = \begin{cases} y, & y \leq M, \\ M, & y > M, \end{cases}$$

whereas quota reinsurance would be written as

$$Ty = (1 - \alpha)y.$$

With this notation we may rewrite Δ_c as

$$\Delta_c = \frac{1}{\kappa} \int_0^\infty (e^{\kappa y} - e^{\kappa Ty}) \, dP(y) - (1 + \eta_r) \int_0^\infty (y - Ty) \, dP(y) \qquad (6.35)$$

and

$$\Delta_r = (\eta_r - \eta_{rc}) \int_0^\infty (y - Ty)\, dP(y), \qquad (6.36)$$

where η_{rc} is the loading required by the reinsurer to attain its chosen probability of ruin. We note that $\Delta_c + \Delta_r$ does not depend on η_r but is a function of Ty and η_{rc}.

Lambert (1963) works out the formulas for excess-loss and quota reinsurance for each of the three possibilities:

1. η_r fixed by competition between various reinsurers. In this case the cedent seeks $d\Delta_c/dT = 0$.

2. The reinsurer knows the type and amount of reinsurance that the cedent will use and can express it as a function of η_r, say $T = f(\eta_r)$. The reinsurer thus finds η_r as the solution of

$$\frac{\partial \Delta_r}{\partial \eta} + \frac{\partial \Delta_r}{\partial T} f'(\eta_r) = 0.$$

3. The cedent and reinsurer agree to maximize $\Delta_c \Delta_r$. This requires the partial derivatives with respect to η_r and T, respectively, to be equated to zero.

Finally we mention that Kahn (1964) points out that the experience rating of an entire portfolio (which we considered in Chapter 3) is essentially analogous to the exchange of insurance company portfolios. Following Wolff (1966), we think of the policyholder as a cedent who "reinsures" a quota α of his outgo variable X with the insurance company for a loading η_1 in addition to the net "reinsurance" premium $\alpha\kappa_1$. The policyholder's aggregate outgo thus becomes $(1 - \alpha)X + \alpha\kappa_1 + \eta_1$. We assume that both policyholder and insurance company have the same quadratic utility function (6.8) so that quota "reinsurance" is optimal.

We now revert to the actual situation and suppose that the insurance company decides to pay as dividend D_+ as well as the aggregate outgo X in return for a premium π_1 from the policyholder. The two forms of the "exchange" can be set out as follows:

Table 6.1

	Insurance Company Pays	Policyholder Pays
Conventional situation	$X + D_+ - \pi_1$	$\pi_1 - D_+$
"Reinsurance" situation	$\alpha X - \alpha\kappa_1 - \eta_1$	$(1 - \alpha)X + \alpha\kappa_1 - \eta_1$

Equating the upper item to the lower in each case results in

$$D_+ = \pi_1 - (1 - \alpha)X - \alpha\kappa_1 - \eta_1. \qquad (6.37)$$

If we now put

$$\eta_1 = (1 - a)\pi_1 - \alpha\kappa_1, \qquad (6.38)$$

we get

$$D_+ = a\pi_1 - bX, \qquad b = 1 - \alpha, \qquad (6.39)$$

which is (3.14), except that we do not exclude the possibility of negative dividends.

REFERENCES

Barrois, T. (1835). "Essai sur l'application du calcul des probabilités aux assurances contre l'incendie." *Mém. Soc. Roy. Sci. Agric. Arts Lille*, **1834**, 85–282.

Bernoulli, D. (1738). "Specimen theoriae novae de mensura sortis." *Comm. Acad. Sci. Imp. Petrop.*, **5**, 175–192. [*Translation* (1954): "Exposition of a new theory on the measurement of risk." *Econometrica*, **22**, 23–36.]

Borch, K. (1960a). "An attempt to determine the optimum amount of stop loss reinsurance." *C.R. XVI Cong. Intern. Actu.*, Brussels, **1**, 597–607.

——— (1960b). "Reciprocal reinsurance treaties seen as a two-person cooperative game." *Skand. Aktu. Tidskr.*, **43**, 29–58.

——— (1960c). "The safety loading of reinsurance premiums." *Skand. Aktu. Tidskr.*, **43**, 163–184.

——— (1961a). "Reciprocal reinsurance treaties." *Astin Bull.*, **1**, 170–191.

——— (1961b). "The utility concept applied to the theory of insurance." *Astin Bull.*, **1**, 245–255.

——— (1962a). "The objectives of an insurance company." *Skand. Aktu. Tidskr.*, **45**, 162–175.

——— (1962b). "A contribution to the theory of reinsurance markets." *Skand. Aktu. Tidskr.*, **45**, 176–189.

——— (1962c). "Equilibrium in a reinsurance market." *Econometrica*, **30**, 424–444.

——— (1963). "Recent developments in economic theory and their application to insurance." *Astin Bull.*, **2**, 322–341.

——— (1964). "The optimal management policy of an insurance company." *Proc. Casualty Actu. Soc.*, **51**, 182–197.

——— (1965). "Una generalización de la teoría del riesgo colectivo." *An. Inst. Actu. Españ.*, **5**, 13–29.

——— (1966a). "La dirección de las compañias de seguros. Un estudio dentro de la teoría económica." *An. Inst. Actu. Españ.*, **6**, 217–230. [An English translation of the foregoing two articles is essentially contained in : (1968) "Dynamic decision problems in an insurance company." *Astin Bull.* **5**, 118–131.

——— (1966b). "Control of a portfolio of insurance contracts." *Astin Bull.*, **4**, 59–71.

De Morgan, A. (1845). "Theory of Probabilities." *Encycl. Metrop.* II, London.

DuMouchel, W. H. (1968) "The Pareto optimality of an *n*-company reinsurance treaty." *Skand. Aktu. Tidskr.*, **51**.

Ferra, C. De (1964). "Considerazioni sulle funzioni di utilità in connessione con la teoria del rischio." *Gior. Ist. Ital. Attu.*, **27**, 51–70.

Finetti, B. De (1942). "Einzelwirtschaftlicher und gemeinwirtschaftlicher Gesichtspunkt in der Frage der Rückversicherung." *Bl. Versich. Math.*, **5**, 307–329.

——— [*Translation*] (1942). "Impostazione individuale e impostazione collettiva del problema della riassicurazione." *Gior. Ist. Ital. Attu.*, **20**, 28–53.]

——— (1957). "Su un 'impostazione alternativa della teoría collettiva del rischio." *Trans XV Intern. Cong. Actu.*, New York, **2**, 433–441.

Hickman, J. C., and D. A. Zahn (1966). "Some mathematical views of risk." *J. Risk Ins.*, **33**, 437–446.

Kahn, P. M. (1961). "Some remarks on a recent paper by Borch." *Astin Bull.*, **1**, 265–272.

——— (1964). "The application of utility theory to group experience rating." *Trans. XVII Intern. Cong. Actu.*, London, **3**, 578–588.

Kopal, Z. (1961). *Numerical Analysis*. Chapman and Hall, London.

Lambert, H. (1963). "Contribution à l'étude du comportement optimum de la cédente et du réassureur dans le cadre de la théorie collective du risque." *Astin Bull.*, **2**, 425–444.

Laplace, P. S. (1812). *Théorie Analytique des Probabilités*. Courcier, Paris.

Mounier, G. J. D. (1894). "Zedelijke premie." *Arch. Verzek.-Wetens.*, **1**, 17–32, 77–99, 145–181.

Ohlin, J. (1968). "On a class of measures of dispersion with application to optimal reinsurance." *Inst. Försäkringsmat. Mat. Statist.*, Stockholm.

Pesonen, E. (1967). "On optimal properties of the stop loss reinsurance." *Astin Bull.*, **4**, 175–176.

Savage, L. J. (1954). *The Foundations of Statistics*. Wiley, New York.

Todhunter, I. (1865). *A History of the Mathematical Theory of Probability*. Macmillan, Cambridge.

Vajda, S. (1963). "Minimum variance reinsurance." *Astin Bull.*, **2**, 257–260.

Welten, C. P. (1964). "Reinsurance optimization by means of utility functions." *Vezek. Arch. (Actu. Stud.)*, **6**, 166–175.

Wolff, K.-H. (1966). *Methoden der Unternehmensforschung im Versicherungswesen*. Springer, Berlin.

APPENDIX A

Renewal Processes

Consider the random sequence of events supposed to occur at the instants $\cdots t_{-1}, t_0, t_1, t_2, \ldots, (t_{i+1} > t_i)$. It is assumed that two or more events cannot occur simultaneously and that the stochastic (point) process thus described is stationary, namely, the joint distribution of the numbers of events in a set of k intervals $(k = 2, 3, \ldots)$ is invariant when the set of intervals is subject to a rigid shift in time. The length of the interval between two of the events is written

$$x_i = t_{i+1} - t_i, \qquad i = \cdots -1, 0, 1, 2, \ldots. \qquad (A.1)$$

[The foregoing description refers to an actual realization of a sequence of events and, in general, we use X_i and T_i for the corresponding random variables.]

If the intervals between the events are *independently and identically distributed*, the sequence is known as a *renewal process*. We shall be concerned only with this type of process in what follows (Cox, 1962; Smith, 1958).

It is obvious that formulas for the distribution of the number of events occurring in a fixed interval beginning with the actual occurrence of an event (namely, with one of the t_i) will, in general, differ from those for an interval of the same length, commencing at a given instant τ_0, say. Let us begin by considering the length of the interval between τ_0 and the occurrence of the nth event following it. Write S_n for the random variable representing this length so that

$$S_n = L_1 + X_2 + X_3 + \cdots + X_n, \qquad (A.2)$$

where $L_1 = T_1 - \tau_0$ is the random variable representing the length of the interval from an arbitrary instant τ_0 to the occurrence of the first event thereafter (sometimes called the forward recurrence time). We write $F(\cdot)$ for the d.f. of each of the X_i and $F_1(\cdot)$ for the d.f. of L_1. If $F_n(\cdot)$ is written for

173

the d.f. of S_n,

$$F_n(\cdot) = F_1(\cdot) * F^{(n-1)*}(\cdot), \qquad n = 1, 2, 3, \ldots, \qquad (A.3)$$

where $*$ stands for the operation of convolution and

$$F^{0*}(x) = \begin{cases} 0, & x < 0, \\ 1, & x \geq 0. \end{cases}$$

Instead of S_n, we may consider the number of events that occur within a given fixed interval t, following the arbitrary instant τ_0. Write $N(t)$ for the random variable representing this number, namely the number of values of S_n $(n = 1, 2, 3, \ldots)$ that do not exceed t. Then

$$N(t) = \max \{n \mid S_n \leq t, S_{n+1} > t\} \qquad (A.4)$$

and

$$\mathfrak{I}\{N(t) > n\} = \mathfrak{I}\{S_n \leq t\} = F_n(t). \qquad (A.5)$$

In the notation used throughout this monograph we thus have

$$\mathfrak{I}\{N(t) = n\} \equiv p_n(t) = F_n(t) - F_{n+1}(t), \qquad (A.6)$$

which emphasizes that the counting process $N(t)$ and the point process S_n are alternative viewpoints of the same process. In particular,

$$p_0(t) = 1 - F_1(t). \qquad (A.7)$$

RELATION BETWEEN $p_0(t)$ AND $F(t)$

There is an important relation connecting $p_0(t)$ and $F(t)$ we now proceed to derive (McFadden, 1962; cited in Chapter 4). Let λ^{-1} be the mean of X_i so that

$$\lambda^{-1} = \int_0^\infty x \, dF(x). \qquad (A.8)$$

Consider the half-open interval $(t_0, t_1]$ of length y ending with the occurrence of an event; suppose that τ_0 has been chosen at random to lie in this interval and that $\mathfrak{F}(\cdot)$, where $\mathfrak{F}(0) = 0$, is the d.f. of Y.

Since the process is stationary, the random variable L_1 is uniformly distributed over an open interval $(0, Y)$ which has a length x with probability $d\,\mathfrak{F}(x)$. Hence the probability that L_1 will assume the value z (i.e., lie between z and $z + dz$) is

$$dF_1(z) = dz \int_z^\infty \frac{d\mathfrak{F}(x)}{x}. \qquad (A.9)$$

We now show that

$$d\mathfrak{F}(x) = \lambda x \, dF(x). \qquad (A.10)$$

The probability that Y will assume the value x (i.e., lie between x and $x + dx$) and that the point τ_0 is at an interval z preceding t_1 is equal to the product

$$d\mathcal{F}(x)\frac{dz}{x}.$$

It is also equal to the probability that there will be at least one event in the interval $(\tau_0 - d\tau_0, \tau_0)$ [which is of order dz and which we write as $c\,dz$] and that the next following interval is of length x [the probability of which is $dF(x)$]. Hence

$$d\mathcal{F}(x)\frac{dz}{x} = c\,dz\,dF(x)$$

or

$$d\mathcal{F}(x) = cx\,dF(x)$$

Integrating from $x = 0$ to $x = \infty$, $c = \lambda$ and (A.10) is verified.

On substituting (A.7) and (A.10) into (A.9), we finally obtain

$$-p_0'(z) = F_1'(z) = \lambda\int_z^\infty dF(x) = \lambda[1 - F(z)]. \tag{A.11}$$

As an example of this relation, suppose that the renewal process degenerates into a Poisson process with hazard λ. Then

$$F(x) = 1 - e^{-\lambda x}$$

and by (A.11)

$$F_1'(z) = \lambda e^{-\lambda z}.$$

Thus, if the density of intervals between events is negative exponential, the beginning of any interval may be chosen randomly. This is, in fact, a necessary and sufficient condition for a renewal process to be a Poisson process.

The probability distribution of the discrete random variable $N(t)$ is given by (A.6). Let us now find the mean and variance of this distribution and derive some asymptotic results for the distribution itself.

THE MEAN AND VARIANCE OF $N(t)$

Write $Z(t)$ and $V(t)$ for the mean and variance, respectively, of the random variable $N(t)$, the number of events during an interval of length t following an arbitrary instant τ_0. Then

$$Z(t) = \sum_{n=1}^\infty n\,p_n(t) = \sum_{n=1}^\infty p_n(t)\sum_{k=1}^n 1 = \sum_{k=1}^\infty \sum_{n=k}^\infty p_n(t)$$

$$= \sum_{k=1}^\infty \sum_{n=k}^\infty [F_n(t) - F_{n+1}(t)] = \sum_{k=1}^\infty F_k(t). \tag{A.12}$$

Now from (A.3)

$$\sum_{k=1}^{\infty} F_k(t) = \int_0^t F_1(t - y)\, dF^{0*}(y) + \sum_{n=1}^{\infty} \int_0^t F_1(t - y)\, dF^{n*}(y)$$

$$= \int_0^t F_1(t - y)\, dU(y), \quad \text{where } U(y) = \sum_{n=0}^{\infty} F^{n*}(y).$$

Using the equivalence of (4.48) and (4.50) we have

$$Z(t) = F_1(t) + \int_0^t Z(t - y)\, dF(y), \tag{A.13}$$

so that (A.12) may be regarded as the solution of the renewal equation (A.13).

We now introduce a special notation for Laplace-Stieltjes transforms. Write, generally,

$$\mathcal{LS}\{f(\cdot)\} = \int_0^{\infty} e^{-sx}\, df(x) \equiv f^*(s).$$

We then have $(n = 1, 2, 3, \ldots)$

$$F_n^*(s) = \int_0^{\infty} e^{-st}\, dF_n(t) = \int_0^{\infty} e^{-st}\, dF^{(n-1)*}(t - y) \int_0^t dF_1(y)$$

$$= \int_0^{\infty} e^{-sy}\, dF_1(y) \int_y^{\infty} e^{-s(t-y)}\, dF^{(n-1)*}(t - y)$$

$$= F_1^*(s)[F^*(s)]^{n-1} \tag{A.14}$$

and

$$Z^*(s) = \int_0^{\infty} e^{-st}\, dZ(t) = \int_0^{\infty} e^{-st}\, dF_1(t) + \int_0^{\infty} e^{-st}\, dZ(t - y) \int_0^t dF(y)$$

$$\text{[from (A.13)]}$$

$$= F_1^*(s) + \int_0^{\infty} e^{-sy}\, dF(y) \int_y^{\infty} e^{-s(t-y)}\, dZ(t - y)$$

$$= F_1^*(s) + F^*(s)\, Z^*(s) = \frac{F_1^*(s)}{1 - F^*(s)} \tag{A.15}$$

Let us write $M(t)$ for the function that $Z(t)$ becomes when $F_1(t) \equiv F(t)$, then (A.15) becomes

$$Z^*(s) = \frac{F_1^*(s)}{1 - F^*(s)} = F_1^*(s)\left\{1 + \frac{F^*(s)}{1 - F^*(s)}\right\} = F_1^*(s) + F_1^*(s) \cdot M^*(s)$$

This relation in L.S.T.'s is equivalent to the integral equation

$$Z(t) = F_1(t) + \int_0^t M(t - y)\, dF(y). \tag{A.16}$$

If, then, the equation

$$M(t) = F(t) + \int_0^t M(t - y)\, dF(y)$$

or its L.S.T.

$$M^*(s) = \frac{F^*(s)}{1 - F^*(s)}$$

can be solved fairly easily, the expression (A.16) for $Z(t)$ will require only numerical integration of known functions.

Turning now to the variance of $N(t)$, we note that

$$\sum_{n=1}^{\infty} n(n+1)\, p_n(t) = \sum_{n=1}^{\infty} p_n(t) \sum_{k=1}^{n} 2k = 2 \sum_{k=1}^{\infty} k \sum_{n=k}^{\infty} p_n(t)$$

$$= 2 \sum_{k=1}^{\infty} k\, F_k(t)$$

so that

$$V(t) \equiv \sum_{n=1}^{\infty} n^2\, p_n(t) - Z^2(t) = \sum_{n=1}^{\infty} [n(n+1) - n]\, p_n(t) - Z^2(t)$$

$$= 2 \sum_{k=1}^{\infty} k\, F_k(t) - Z(t)[1 + Z(t)]. \tag{A.17}$$

We evaluate the first term of (A.17) by means of L.S.T.'s. Write

$$G(t) \equiv \sum_{k=1}^{\infty} k\, F_k(t).$$

Then

$$G^*(s) = \int_0^{\infty} e^{-st}\, dG(t) = \sum_{k=1}^{\infty} k \int_0^{\infty} e^{-st}\, dF_k(t)$$

$$= F_1^*(s) \sum_{k=1}^{\infty} k[F^*(s)]^{k-1}, \qquad \text{from (A.14)}$$

$$= \frac{F_1^*(s)}{[1 - F^*(s)]^2} = \frac{F_1^*(s)\, F(s)}{[1 - F^*(s)]^2} + \frac{F_1^*(s)}{1 - F^*(s)}$$

$$= Z^*(s)M^*(s) + Z^*(s).$$

Hence on inversion

$$G(t) = \int_0^t Z(t - y)\, dM(y) + Z(t)$$

and on substitution into (A.17)

$$V(t) = 2 \int_0^t Z(t - y)\, dM(y) + Z(t)[1 - Z(t)]. \tag{A.18}$$

ASYMPTOTIC OR "EQUILIBRIUM" RESULTS

Relation 4.55 shows that for large t

$$M(t) \sim \frac{t}{\displaystyle\int_0^{\infty} y\, dF(y)} \sim Z(t), \tag{A.19}$$

but we need to derive an asymptotic relation for $V(t)$. Now (A.19) implies that $Z^*(s) \sim 1/s\mu$, where $\mu = \int_0^\infty h \, dF(y)$, and the L.S.T. of $\int_0^t M(y) \, dy$ is $s^{-1}M^*(y)$. Thus, corresponding to the relation

$$G^*(s) = Z^*(s) + Z^*(s)M^*(s),$$

there is an asymptotic inversion

$$G(t) \sim \frac{t}{\mu} + \frac{1}{\mu} \int_0^t M(y) \, dy.$$

Hence (A.18) provides

$$V(t) \sim 2\left[\frac{t}{\mu} + \frac{1}{\mu}\int_0^t M(y) \, dy\right] - \frac{t}{\mu} - \frac{t^2}{\mu^2}$$

$$= \frac{t}{\mu} - \frac{t^2}{\mu^2} + \frac{2}{\mu}\int_0^t M(y) \, dy. \tag{A.20}$$

Now (A.5) shows that $-N(t)$ has the same d.f. as the sum of n independent random variables and, by the Central Limit theorem, this d.f. is asymptotically Normal. Hence

$$\mathfrak{F}\{-N(t) \le -n\} = \mathfrak{F}\{N(t) > n\} \sim \Phi(z), \tag{A.21}$$

where

$$z = \frac{-n - [-Z(t)]}{V^{1/2}(t)} = \frac{Z(t) - n}{V^{1/2}(t)}.$$

Since t must be large if n is to be large, we may use relations (A.19) and (A.20) for $Z(\cdot)$ and $V(\cdot)$ in (A.21).

On the other hand, if σ^2, the variance of $F(\cdot)$, is readily available, we may prefer to write

$$\mathfrak{F}\{S_n \le t\} \equiv \mathfrak{F}\left\{\frac{S_n - n\mu}{\sigma\sqrt{n}} < \frac{z}{(1 - \sigma z/\sqrt{t\mu})^{1/2}}\right\},$$

where

$$n = \frac{t}{\mu} - \sigma z\left(\frac{t}{\mu^3}\right)^{1/2};$$

S_n is asymptotically Normally distributed with mean $n\mu$ and variance $n\sigma^2$ so that when n and t both tend to infinity with z fixed

$$\mathfrak{F}\{S_n \le t\} \sim \Phi(z), \tag{A.22}$$

where

$$Z = \frac{t/\mu - n}{\sigma\sqrt{t/\mu^3}}.$$

Relations A.21 and A.22 are essentially in agreement when the right-hand side of (A.20) is given the value t/μ^3.

Illustration. Carleton (1945) has considered the situation in which the intervals between accidents occurring to an automobile (say) would follow a Poisson process with hazard λ (so that their density would be negative exponential) except that each accident is followed by a period of repair during which no accident can occur. For simplicity it was assumed that the duration of repair was a constant equal to h. Zero time was assumed to occur when the automobile was in operation.

This problem is now known as that of the Type I Geiger-Müller counter and has a history going back to 1913 (Smith, 1958). Although Morant (1921) derived an expression for $p_n(t)$, it appears to be incorrect, and it was not until 1936 that Ruark and Devol obtained the result that Carlton was later to derive independently. We note that if the single automobile were replaced by a large fleet with an aggregate hazard λ and only a single repair shop were available we would have Erlang's original (1909) queueing problem discussed in Chapter 4.

Viewed as a renewal process, the above accident situation implies that

$$F_1(t) = 1 - e^{-\lambda t}, \qquad 0 \le t < \infty,$$

and

$$F(t) = 1 - e^{-\lambda(t-h)}, \qquad h \le t < \infty.$$

[We note that in the more general case in which the period of repair is a random variable with d.f. $P(\cdot)$

$$F(t) = \int_0^t P(z)\lambda e^{-\lambda(t-z)}\, dz$$

Useful comments on this relation are found in Smith (1958).]

Thus

$$F_1^*(s) = \lambda \int_0^\infty e^{-(s+\lambda)t}\, dt = \frac{\lambda}{\lambda + s}$$

and

$$F^*(s) = \lambda \int_h^\infty e^{-st} e^{-\lambda(t-h)}\, dt = \lambda e^{-sh} \int_0^\infty e^{-(s+\lambda)\tau}\, d\tau = \frac{\lambda e^{-sh}}{\lambda + s}.$$

Relation A.14 then provides

$$F_n^*(s) = \frac{\lambda}{\lambda + s}\left(\frac{\lambda e^{-sh}}{\lambda + s}\right)^{n-1} = e^{-(n-1)sh}\frac{\lambda^n}{(\lambda + s)^n}, \qquad n = 1, 2, 3, \ldots,$$

and it is easily verified that the inverse of this L.S.T. is zero when $t \le (n-1)h$ and

$$F_n(t) = \frac{\lambda^n}{\Gamma(n)}\int_{(n-1)h}^t (\tau - \overline{n-1}h)^{n-1}e^{-\lambda(\tau-\overline{n-1}h)}\, d\tau, \qquad t \ge (n-1)h,$$

$$= \frac{1}{\Gamma(n)}\int_0^{\lambda(t-\overline{n-1}h)} z^{n-1}e^{-z}\, dz$$

$$= e^{-\lambda(t-\overline{n-1}h)}\sum_{j=n}^\infty \frac{\lambda^j(t-\overline{n-1}h)^j}{j!}. \qquad \text{by repeated integration}$$

Hence, from (A.6), with $t \geq (n-1)h$ and $n = 1, 2, 3, \ldots$,

$$p_n(t) = F_n(t) - F_{n+1}(t) = [1 - F_{n+1}(t)] - [1 - F_n(t)]$$

$$= e^{-\lambda(t-nh)} \sum_{j=0}^{n} \frac{\lambda^j(t-nh)^j}{j!} - e^{-\lambda(t-\overline{n-1}h)} \sum_{j=0}^{n-1} \frac{\lambda^j(t-\overline{n-1}h)^j}{j!}$$

in agreement with Carleton (1945). The two terms of this result could, of course, be written in terms of incomplete gamma functions. For any given value of $t > 0$ we have

$$\sum_{n=1}^{\infty} p_n(t) = \sum_{n=1}^{\infty} [F_n(t) - F_{n+1}(t)] = F_1(t),$$

so that

$$p_0(t) = 1 - F_1(t) = e^{-\lambda t},$$

as we know from the terms of the problem.

REFERENCES

Carleton, J. (1945). "Non-random accident distributions and the Poisson series." *Proc. Casualty Actu. Soc.*, **32**, 21–26.

Cox, D. R. (1962). *Renewal Theory*. Methuen, London.

Morant, G. (1921). "On random occurrences in space and time, when followed by a closed interval." *Biometrika*, **13**, 309–337.

Ruark, A., and L. Devol (1936). "The general theory of fluctuations in radioactive disintegration." *Phys. Rev.*, **49**, 355–367.

Smith, W. L. (1958). "Renewal theory and its ramifications." *J. Roy. Statist. Soc.*, B. **20**, 243–302.

APPENDIX B

Numerical Inversion of Laplace Transforms

The Laplace transform (L.T.) $\phi(s)$ of a function $f(z)$, $0 < z < \infty$, is defined by

$$\phi(s) = \int_0^\infty e^{-sz} f(z)\, dz, \qquad \Re s > 0,\ \lim_{z \to \infty} e^{-sz} f(z) = 0.$$

We suppose that $\phi(s)$ can be calculated numerically for any given set of real positive values of s and that we wish to find approximate values for $f(z)$ at a number of selected values of z. We observe that if an infinite value of $\phi(0)$ would invalidate any method of inversion we could work, instead, with $\phi(s) - s^{-1} \lim_{s \to \infty} [s\phi(s)]$ when there is a simple pole or, alternatively, with $(1 - e^{-Ts})\, \phi(s)$, where T is very large. In the first case the inverse has to be increased by a constant, whereas in the second it has to be increased by $f(z - T)$ only when $z > T$; T should thus be chosen beyond the range of z-values in which we are interested.

The saddlepoint approximations of (2.47) have already provided a method of inversion of $\phi(s) = e^{K(s)}$, which depends for its accuracy on $K''(u)$ being large where u, a real positive number, is given by (2.46), namely,

$$K'(u) = \frac{\phi'(u)}{\phi(u)} = x;$$

but the literature, most of it very recent, describes four other essentially different methods of effecting this numerical inversion of $\phi(s)$, and emphasis is frequently laid on the fact that the success of the method used will depend on the behavior of $f(z)$ as well as on the form of $\phi(s)$. Compared with many physical problems, our $f(\cdot)$ is likely to assume a relatively simple shape, since it is usually a monotonically increasing probability or a well-behaved density function. Since we have not found any numerical examples in the actuarial or statistical literature, we shall describe all four methods sufficiently for them to be applied in practice.

181

$f(z)$ A POLYNOMIAL IN SOME FUNCTION OF z

The problem here is to choose x as some function of z that transforms $f(z)$ into a function $g(x)$, say, which can be expressed as a polynomial in x and will also lead to simple numerical calculations.

As an example of the procedure, let us write

$$x = 1 - e^{-z/a}, \qquad 0 < x < 1,$$

and substitute it into the integrand of $\phi(s)$. Without difficulty we obtain

$$\phi(s) = a \int_0^1 (1 - x)^{as-1} f(-a \ln \overline{1 - x}) \, dx \tag{B.1}$$

and for convenience we write

$$f(-a\ln \overline{1 - x}) \equiv g(x),$$

where $g(x)$ is to be expanded as a polynomial in x. Instead of equating $g(x)$ to a standard polynomial in x with undetermined coefficients, we utilize the Jacobi orthogonal polynomials defined on an interval $(0, 1)$ [Hochstrasser (1964); Magnus et al. (1966); cited in Chapter 5]. We thus write

$$g(x) = \sum_{n=0}^{\infty} C_n \, G_n(p, q, x), \tag{B.2}$$

where $G_n(p, q, x)$, $q > 0$, $p - q > -1$, is the Jacobi polynomial in x of degree n, which satisfies the orthogonality relation

$$
\int_0^1 x^{q-1}(1 - x)^{p-q} \, G_n(p, q, x) \, G_m(p, q, x) \, dx
$$
$$
= \begin{cases} 0, & m \neq n, \\ n! \, \dfrac{\Gamma(p + n) \, \Gamma(q + n) \, \Gamma(p - q + n + 1)}{(2n + p)[\Gamma(p + 2n)]^2} & m = n. \end{cases}
$$

For reasons that will appear, we choose $q = 1$ and write $s = (p + k)/a$, where $k = 0, 1, 2, 3, \ldots$. Inserting (B.2) in (B.1), we then have

$$\phi\left(\frac{p + k}{a}\right) = a \int_0^1 (1 - x)^{p+k-1} \left[\sum_{n=0}^{\infty} C_n \, G_n(p, 1, x) \right] dx$$

$$= a \sum_{n=0}^{\infty} C_n \int_0^1 (1 - x)^{p+k-1} \, G_n(p, 1, x) \, dx. \tag{B.3}$$

Now, with p and a suitably chosen, we may calculate the left-hand side of (B.3) for $k = 0, 1, 2, \ldots$. Our objective is to evaluate the integral on the right of (B.3). For this purpose we write

$$(1 - x)^k \equiv \sum_{j=0}^{k} a_j\, G_j(p, 1, x), \qquad k = 0, 1, 2, \ldots,$$

so that

$$\int_0^1 (1 - x)^{p+k-1}\, G_n(p, 1, x)\, dx = \sum_{j=0}^{k} a_j \int_0^1 (1 - x)^{p-1}\, G_n(p, 1, x)\, G_j(p, 1, x)\, dx$$

$$= \begin{cases} 0, & k < n, \\[2mm] a_n\, \dfrac{(n!)^2\, \Gamma^2(p + n)}{(2n + p)\, \Gamma^2(p + 2n)} & k \geq n, \end{cases} \tag{B.4}$$

by means of the orthogonality relation. On the other hand, if we write

$$G_n(p, 1, x) \equiv \sum_{j=0}^{n} b_j (1 - x)^j, \tag{B.5}$$

then

$$\int_0^1 (1 - x)^{p+k-1}\, G_n(p, 1, x)\, dx = \sum_{j=0}^{n} b_j \int_0^1 (1 - x)^{p+k+j-1}\, dx$$

$$= \sum_{j=0}^{n} \frac{b_j}{p + k + j} = \frac{Q_n(k)}{\displaystyle\prod_{j=0}^{n} (p + k + j)}, \tag{B.6}$$

where $Q_n(k)$ is a polynomial in k of degree n; but we have just seen that the integral vanishes for $k = 0, 1, 2, \ldots, n - 1$, so that

$$Q_n(k) = A k^{(n)}, \qquad k \geq n,$$

where A is a constant. Now when $k = n = 0$, (B.6) provides

$$b_0 = A.$$

On the other hand (Hochstrasser, 1964),

$$G_n(p, 1, 0) = (-1)^n\, \frac{n!\, \Gamma(p + n)}{\Gamma(p + 2n)} = \sum_{j=0}^{n} b_j \quad \text{by (B.5)}$$

and thus

$$G_0(p, 1, 0) = 1 = b_0.$$

Therefore (B.4) and (B.6) finally result in

$$\int_0^1 (1 - x)^{p+k-1}\, G_n(p, 1, x)\, dx = \frac{k^{(n)}\, \Gamma(p + k)}{\Gamma(p + k + n + 1)}, \qquad k \geq n,$$

and (B.3) becomes $(k = 0, 1, 2, \ldots)$

$$\phi\!\left(\frac{p + k}{a}\right) = a \sum_{n=0}^{k} \frac{k^{(n)}\, \Gamma(p + k)}{\Gamma(p + k + n + 1)}\, C_n. \tag{B.7}$$

The coefficients C_n are determined recursively from (B.7) with $k = 0, 1, 2, \ldots$ in succession. Then from (B.2) we obtain

$$g(x) = f(-a \ln \overline{1 - x}) = \sum_{n=0}^{\infty} C_n G_n(p, 1, x), \qquad \text{(B.8)}$$

where $x = 1 - e^{-z/a}$ and the lower order Jacobi polynomials are available in tables and beyond $n = 5$ or 6 can be calculated individually (Hochstrasser, 1964).

The foregoing development in terms of Jacobi polynomials is due to Miller and Guy (1966), who give six numerical illustrations, each with 10 or 11 terms in the series. They recommend that p should generally be chosen between 0.5 and 6.0 and a between 0.5 and 20.0, with the larger p- and a-values for small values of z.

Expansions in terms of Legendre orthogonal polynomials have been proposed by Erdélyi (1943), Papoulis (1956), and Lanczos (1956). Following the latter author, we write $x = e^{-z/a}$ and suppose we have calculated

$$\phi\left(\frac{k + 1}{a}\right) = a \int_0^1 x^k g(x)\, dx, \qquad k = 0, 1, 2, \ldots,$$

where $g(x) = f(-a \ln x)$. Assume that $g(\cdot)$ can be expanded in a series of Legendre polynomials shifted to the range (0, 1)* instead of the customary $(-1, 1)$. We write the corresponding polynomial of the kth degree as $P_k^*(x)$ and put

$$f(-a \ln x) = g(x) = \sum_{j=0}^{N}(2j + 1)c_j P_j^*(x), \qquad \text{(B.9)}$$

where (Hochstrasser, 1964)

$$\int_0^1 \{P_j^*(x)\}^2\, dx = \frac{1}{2j + 1}.$$

Multiplying (B.9) by $P_k^*(x)$ and integrating over the range of orthogonality, we have

$$\int_0^1 P_k^*(x)g(x)\, dx = \sum_{j=0}^{N}(2j + 1)c_j \int_0^1 P_k^*(x)P_j^*(x)\, dx$$
$$= c_k,$$

and thus, if the coefficient of x^l in $P_k^*(x)$ is b_{kl},

$$c_k = \int_0^1 \left(\sum_{l=0}^{k} b_{kl}x^l\right)g(x)\, dx = a^{-1}\sum_{l=0}^{k} b_{kl}\phi\left(\frac{l + 1}{a}\right). \qquad \text{(B.10)}$$

The coefficients of the series expansion (B.9) are thus calculated numerically by means of (B.10). The parameter a is to be chosen as before.

---

* Lanczos provides the exact coefficients of the first 15 of these polynomials in his Table IV.

Papoulis (1956) and Lanczos (1956) have also provided expressions in terms of Chebyshev and Laguerre orthogonal polynomials. Suppose we arrange that $f(z) = 0$ [e.g., by subtracting the actual value of $f(0)$] and replace (B.10) with

$$f(-a \ln x) = g(x) = \frac{8}{\pi} \sum_{j=0}^{N} c_j (x - x^2)^{1/2} U_j^*(x), \qquad (\text{B.11})$$

where $U_j^*(x)$ is a Chebyshev polynomial of the second kind shifted to the range $(0, 1)$. This polynomial has a weight function $(x - x^2)^{1/2}$ and satisfies

$$\int_0^1 (x - x^2)^{1/2} \{U_j^*(x)\}^2 \, dx = \frac{\pi}{8} \qquad (\text{Hochstrasser, 1964})$$

Then, exactly as before,

$$c_k = a^{-1} \sum_{l=0}^{k} b_{kl} \phi \left(\frac{l+1}{a} \right), \qquad (\text{B.12})$$

where b_{kl} is now the coefficient of x^l in $U_k^*(x)$ and is tabulated by Lanczos (1956) for values of k up to 13.

Last we write $f(z)$ in the form

$$f(z) = \sum_{n=0}^{\infty} C_n e^{-az} L_n(z),$$

where $L_n(z)$ is the Laguerre orthogonal polynomial defined on the interval $(0, \infty)$ and given by

$$L_n(z) = \frac{e^z}{n!} \frac{d^n}{dz^n} (e^{-z} x^n) = \sum_{j=0}^{n} (-1)^j \binom{n}{j} \frac{z^j}{j!}.$$

This type of expansion has been proposed for probability density functions by Khamis (1960) and illustrated by Bowers (1966). The Laplace transform in this case is

$$\phi(s) = \sum_{n=0}^{\infty} C_n \int_0^{\infty} e^{-(s+a)z} \sum_{j=0}^{n} (-1)^j \binom{n}{j} \frac{z^j}{j!} \, dz$$

$$= \sum_{n=0}^{\infty} C_n \sum_{j=0}^{n} (-1)^j \binom{n}{j} (s+a)^{-j-1} = \sum_{n=0}^{\infty} C_n \frac{(s+a-1)^n}{(s+a)^{n+1}} ; \quad (\text{B.13})$$

but

$$\phi(s) = \phi(\overline{1-a} + \overline{s+a-1}) = \sum_{k=0}^{\infty} \frac{(s+a-1)^k}{k!} \phi^{(k)}(1-a)$$

$$= \sum_{k=0}^{\infty} \frac{\phi^{(k)}(1-a)}{k!} \frac{1}{s+a} \left(1 - \frac{1}{s+a}\right)^k \left[1 - \left(1 - \frac{1}{s+a}\right)\right]^{-k+1}$$

$$= \frac{1}{s+a} \sum_{k=0}^{\infty} \frac{\phi^{(k)}(1-a)}{k!} \sum_{n=k}^{\infty} \binom{n}{k} \left(1 - \frac{1}{s+a}\right)^n$$

$$= \frac{1}{s+a} \sum_{n=0}^{\infty} \left(1 - \frac{1}{s+a}\right)^n \sum_{k=0}^{n} \binom{n}{k} \frac{\phi^{(k)}(1-a)}{k!},$$

and comparison with (B.13) shows that

$$C_n = \sum_{k=0}^{n} \binom{n}{k} \frac{\phi^{(k)}(1-a)}{k!}.$$

Hence

$$f(z) = \sum_{n=0}^{\infty} e^{-az} L_n(z) \sum_{k=0}^{n} \binom{n}{k} \frac{\phi^{(k)}(1-a)}{k!}, \qquad (B.14)$$

where the derivatives of $\phi(\cdot)$ at the argument $1-a$ are to be calculated by standard numerical methods (Davis and Polonsky, 1964). This particular method of inversion of a Laplace transform actually goes back to the first half of the nineteenth century (Bateman, 1936).

This review of L.T. inversion by expressing $f(z)$ as a polynomial in a function of z is concluded by mentioning that the transformation

$$1 - x^2 = e^{-z/a}$$

leads to a series in the ultraspherical (Gegenbauer) orthogonal polynomials of odd degree. Berger (1966) has actually applied this transformation to the two-dimensional Laplace transform

$$\phi(r, s) = \int_0^{\infty} \int_0^{\infty} e^{-rz_1 - sz_2} f(z_1, z_2)\, dz_1\, dz_2 \qquad (B.15)$$

and has obtained $(k_1, k_2 = 0, 1, 2, \ldots)$

$$\phi\left(\frac{k_1 + \lambda_1 + \frac{1}{2}}{a_1}, \frac{k_2 + \lambda_2 + \frac{1}{2}}{a_2}\right)$$

$$= a_1 a_2 \sum_{n_1=0}^{k_1} \sum_{n_2=0}^{k_2} C_{n_1 n_2} A(\lambda_1, k_1, 2n_1 + 1) A(\lambda_2, k_2, 2n_2 + 1), \quad (B.16)$$

where

$$A(\lambda, k, 2n + 1) = \frac{(-1)^n \Gamma(\frac{1}{2}) \Gamma(k + \lambda + \frac{1}{2}) \Gamma(n + \lambda + 1)k!}{\Gamma(\lambda) \Gamma(n + k + \lambda + 2)n!\,(k-n)!}.$$

The coefficients $C_{n_1 n_2}$ are computed recursively and we then have

$$g(x_1, x_2) = f(-a_1 \ln \overline{1 - z_1^2}, -a_2 \ln \overline{1 - z_2^2})$$

$$= \sum_{n_1=0}^{\infty} \sum_{n_2=0}^{\infty} C_{n_1 n_2} C_{2n_1+1}^{(\lambda_1)}(x_1) C_{2n_2+1}^{(\lambda_2)}(x_2), \qquad (B.17)$$

where $C_{2n+1}^{(\lambda_1)}(x)$ is the ultraspherical polynomial of degree $2n + 1$ (Hochstrasser, 1964). Two short (one-dimensional) numerical illustrations are included in the article.

In spite of the relative simplicity of the foregoing methods of inversion, they are not suitable for calculations on a desk machine. The coefficients C_n change sign irregularly and it is necessary to work to many significant figures. Furthermore, the extension of available tables of orthogonal polynomials to more than order six is laborious. In general, double precision computer routines are necessary.

$\phi(s)$ A POLYNOMIAL IN s^{-1}

It is clear from its definition that $\phi(s)$ decreases as $\Re s$ increases. It is thus natural to suppose that $\phi(s)$ may be capable of representation as a polynomial in s^{-1} without a constant term. Write

$$\phi(s) = \sum_{n=1}^{m} B_n s^{-n}$$

and express this as the $(m + 1)$-point Lagrange polynomial in s^{-1} based on $\phi(s)$ at $s = 1, 2, 3, \ldots, m$ and $s = \infty$. Then

$$\phi(s) = \sum_{j=1}^{m} \frac{(1/s) \prod\limits_{j \neq k=1}^{m} (1/s - 1/k)}{(1/j) \prod\limits_{j \neq k=1}^{m} (1/j - 1/k)} \phi(j)$$

$$= \sum_{j=1}^{m} \frac{j^{m+1}}{s^{m+1}} \frac{\prod\limits_{j \neq k=0}^{m} (s - k)}{\prod\limits_{j \neq k=0}^{m} (j - k)} \phi(j) = \sum_{j=1}^{m} \frac{j^{m+1}}{s^{m+1}} Q_{mj}(s)\, \phi(j),$$

where the products range over the values of k indicated, the $(m + 1)$th term is missing because $\phi(\infty) = b$, and $Q_{mj}(s)$ is a polynomial in s of degree m without a constant term. Writing $Q_{mj}(s) = \sum_{k=1}^{m} c_{jk} s^k$ and inverting

$$f(z) = \mathcal{L}^{-1}[\phi(s)] = \sum_{j=1}^{m} j^{m+1} \phi(j) \sum_{k=1}^{m} \frac{z^{m-k}}{(m-k)!} c_{jk}, \qquad (B.18)$$

where c_{jk} is obtained from

$$\frac{s(s-1)(s-2)\cdots(s-\overline{k-1})(s-\overline{k+1})\cdots(s-m)}{j(j-1)(j-2)\cdots(j-\overline{k-1})(j-\overline{k+1})\cdots(j-m)} = \sum_{k=1}^{m} c_{jk} s^k. \quad (B.19)$$

This straightforward method of calculating $f(z)$ in terms of m equidistant values of $\phi(s)$ was suggested by Salzer (1958), who has provided extensive tables of

$$A_j^{(m)}(z) = j^{m+1} \sum_{k=1}^{m} \frac{z^{m-k}}{(m-k)!} c_{jk}$$

for $m = 1(1)6$ and $z = 0.0(0.1)6.0$; $m = 7, 8$ and $z = 0.0(0.2)m$; $m = 9, 10$ and $z = 0(1)m$; and has illustrated them by means of five numerical examples. These examples strikingly illustrate the loss of significant figures in the calculations. Although eight or nine figures are used throughout, only four may be left in $f(\cdot)$. Here again it would seem that double precision computer calculations are desirable.

n VALUES OF $\phi(s)$ AS LINEAR RELATIONS
IN CHOSEN VALUES OF $f(z)$

Once again we write $x = e^{-z/a}$ and obtain

$$\phi(s) = a \int_0^1 x^{as-1} g(x)\, dx, \qquad (B.20)$$

where

$$g(x) = f(-a \ln x).$$

We could use any quadrature formula to reduce the integral to the sum of n terms (plus a remainder which we ignore). Substantial simplifications occur if we utilize the n values of x which correspond to the zeros of an orthogonal polynomial of nth degree. This is known as Gaussian quadrature and possesses the considerable advantage that the use of n ordinates produces an area that is correct for a polynomial of any degree up to $2n - 1$ (Lanczos, 1956).

Let us write $Q_n(x)$ for a polynomial of degree n that is orthogonal over the interval $(0, 1)$* for a weight function $w(x)$, namely, $(m, n = 0, 1, 2, \ldots)$

$$\int_0^1 w(x)\, Q_n(x)\, Q_m(x)\, dx = \begin{cases} 0, & m \neq n, \\ h_n, & m = n, \end{cases}$$

where $h_n^{1/2}$ is the normalization factor. It follows that if $\Pi_m(x)$ is any polynomial of degree $m < n$

$$\int_0^1 w(x)\, \Pi_m(x)\, Q_n(x)\, dx = 0$$

and that all the n zeros of $Q_n(x)$ are simple and located in $(0, 1)$.

By using these and other derived properties of $Q_n(x)$, it can be shown (Stroud and Secrist, 1966) that the relation

$$\int_0^1 k(x)\, dx = \int_0^1 w(x)\, h(x)\, dx = \sum_{j=1}^n A_j\, h(x_j) = \sum_{j=1}^n \frac{A_j}{w(x_j)}\, k(x_j) \quad (B.21)$$

will hold exactly for $h(x)$ a polynomial of degree $2n - 1$ or less if

$$A_j = \int_0^1 w(x)\, \frac{Q_n(x)}{(x - x_j)\, Q_n'(x)}\, dx = \frac{h_{n-1}}{Q_n'(x_j)\, Q_{n-1}(x_j)} \qquad (B.22)$$

and x_j $(j = 1, 2, \ldots, n)$ are the n real roots of $Q_n(x) = 0$.

Reverting to (B.20), writing $s = (k + 1)/a$, $k = 0, 1, 2, \ldots, n - 1$, and using (B.21), we obtain

$$a^{-1} \phi\!\left(\frac{k + 1}{a}\right) = \int_0^1 x^k\, g(x)\, dx = \sum_{j=1}^n \frac{A_j}{w(x_j)}\, x_j^k\, g(x_j). \qquad (B.23)$$

---

*Note that polynomials orthogonal over any finite interval (a, b) can be "normalized" to $(0, 1)$.

This is a system of n equations in the n unknowns $g(x_j)$, namely, $f(-a \ln x_j)$. For brevity we write the system as

$$\sum_{j=1}^{n} x_j^k y_j = a_k, \qquad k = 0, 1, 2, \ldots, n - 1.$$

Now multiply through equation k by an arbitrary constant q_k and sum the n results thus obtained, namely,

$$\sum_{j=1}^{n} y_j \left(\sum_{k=0}^{n-1} q_k x_j^k \right) = \sum_{k=0}^{n-1} a_k q_k. \qquad (B.24)$$

The sum in parentheses is then an arbitrary polynomial in x_j of degree $n - 1$. Let us write, generally,

$$\sum_{k=0}^{n-1} q_{kj} x^k \equiv \frac{Q_n(x)}{(x - x_j) Q_n'(x_j)}, \qquad (B.25)$$

where the $(n - 1)$th degree polynomial in x on the right clearly has the property of equaling zero when $x \neq x_j$ and of equaling unity when $x = x_j$. Since $Q_n(x)$ is supposed to be a suitably chosen family of orthogonal polynomials [namely, that leading to the A_j of (B.23)], the q_{kj} ($j = 1, 2, \ldots, n; k = 0, 1, \ldots, n - 1$) are uniquely determined. With $x = x_l$ in (B.24), the use of (B.25) implies

$$y_l = \sum_{k=0}^{n-1} a_k q_{kl}$$

or, in our original notation ($l = 1, 2, 3, \ldots, n$),

$$g(x_l) = f(-a \ln x_l) = \frac{w(x_l)}{a A_l} \sum_{k=0}^{n-1} q_{kl}\, \phi\!\left(\frac{k + 1}{a}\right), \qquad (B.26)$$

where the n coefficients q corresponding to x_l are determined by substituting x_l for x_j in (B.25).

The two relations (B.25) and (B.26) thus provide the means of calculating n unequally spaced values of $f(z)$ corresponding to n equally spaced values of $\phi(s)$. To facilitate the computations it is useful to have available tables of the weights $w(x_j)$ at the n zeros of the chosen family $Q_n(x)$, the Gaussian quadrature coefficients A_j (both of which are provided by Stroud and Secrist, 1966), and the n coefficients of the $(n - 1)$th degree polynomial (B.25). The Jacobi and Gegenbauer polynomials pose problems in that they require the choice of values for the parameters (α, β) and λ, respectively—although, of course, this adds to the flexibility of the inversion. The only thorough application of the method has been that of Bellman et al. (1966), who utilized the Legendre polynomials $P_n(x)$ "shifted" from $(-1, 1)$ to $(0, 1)$. Many numerical examples are given in the last three of the book's five chapters and the seventeen digit tables appended thereto indicate the impracticability of a desk machine limited to eight or nine significant figures.

$\phi(s)$ EXPRESSED AS THE QUOTIENT OF TWO POLYNOMIALS

Suppose that $\phi(s)$ can be expressed in the form

$$\phi(s) = \frac{\displaystyle\sum_{j=1}^{n} a_j s^{n-j}}{s^n + \displaystyle\sum_{j=1}^{n} b_j s^{n-j}} = \frac{\displaystyle\sum_{j=1}^{n} a_j s^{-j}}{1 + \displaystyle\sum_{j=1}^{n} b_j s^{-j}} \equiv \frac{\pi(s)}{1 + \chi(s)}, \qquad (B.27)$$

where a_j and b_j $(j = 1, 2, \ldots, n)$ are known. By writing

$$\mathcal{L}^{-1}\{\pi(s)\} = p(x) = \sum_{j=1}^{n} a_j \frac{x^{j-1}}{(j-1)!} \quad \text{and} \quad \mathcal{L}^{-1}\{\chi(s)\} = q(x) = \sum_{j=1}^{n} \frac{b_j x^{j-1}}{(j-1)!}$$

$$(B.28)$$

the relation $\phi(s) + \phi(s)\chi(s) = \pi(s)$ can be inverted to become

$$f(x) + \int_0^x f(y)\, q(x - y)\, dy = p(x). \qquad (B.29)$$

Numerical values of $p(x)$ and $q(x)$ are now calculated at equidistant values $l\delta$ $(l = 0, 1, 2, \ldots)$; write the results as p_k and q_k, respectively. Let us suppose, for the moment, that f_0 and f_1 have been calculated. If l is even, say $2k$ $(k = 1, 2, 3, \ldots)$, we apply the repeated Simpson quadrature formula to the integral in (B.29) and solve for f_{2k}. The result is

$$f_{2k} = \frac{\begin{aligned}p_{2k} - (\delta/3)[f_0 q_{2k} + 4(f_1 q_{2k-1} + f_3 q_{2k-3} + \cdots) \\ + 2(f_2 q_{2k-2} + f_4 q_{2k-4} \cdots)]\end{aligned}}{1 + \delta q_0/3}, \qquad (B.30)$$

where the term with the coefficient 2 is omitted when $k = 1$. On the other hand, if l is odd, say $2k + 1$ $(k = 1, 2, 3, \ldots)$, we apply the three-eighths rule over the interval $(0, \delta)$ and the repeated Simpson (when $k = 2, 3, \ldots$) over the interval $(3\delta, 2k + 1\, \delta)$. On solving for f_{2k+1}, the result is

$$f_{2k+1} = \frac{\begin{aligned}p_{2k+1} - (3\delta/8)(f_0 q_{2k+1} + 3f_1 q_{2k} + 3f_2 q_{2k-1} + \tfrac{17}{9} f_3 q_{2k-2}) \\ - (\delta/3)[4(f_4 q_{2k-3} + \cdots) + 2(f_5 q_{2k-4} + \cdots)]\end{aligned}}{1 - \delta q_0/3} \qquad (B.31)$$

except that when $k = 1$ the coefficient of the last term in the first square bracket is unity instead of $\tfrac{17}{9}$.

It will be seen that when f_0 and f_1 are assumed to be known, insertion of $k = 1$ in (B.30) will provide f_2; then use of $k = 1$ in (B.31), as modified, will provide f_3, and so on, using (B.30) and (B.31) alternately.

In order to calculate the boundary value f_0, we observe that

$$f_0 = f(0) = \lim_{s \to 0} s\, \phi(s) = a_1,$$

which is assumed known. Then to obtain a first approximation to f_1, say f_1^*, we use the trapezoidal rule on the integral in (B.29) so that f_1^* is obtained from

$$f_1^* + \frac{\delta}{2}(f_0 q_1 + f_1^* q_0) = p_1.$$

This first approximation is then used to calculate f_2, say f_2^*, from (B.30), as modified, with $k = 1$. We then apply the Lagrangian integration formula (Davis and Polonsky, 1964)

$$\int_0^1 g(x)\,dx = \tfrac{1}{12}[5\,g(0) + 8\,g(1) - g(2)],$$

which, when used on (B.29), gives

$$f_1 + \frac{\delta}{12}(5f_0 q_1 + 8f_1 q_0 - f_2^* q_{-1}) = p_1.$$

In this equation $q_{-1} = q(-\delta)$ is calculated from (B.28) and the new approximation to f_1 is obtained. A substitution of this value into (B.30) with $k = 1$ produces a second approximation for f_2. This procedure could, of course, be repeated more than once if f_1^* and f_1 differed greatly.

Once a whole set of values of $\{f_l\}$ has been calculated for a given value of δ, this value may be halved and the whole process repeated. The use of formulas (B.30) and (B.31) is very rapid up to quite large l-values on an electronic computer, and several δ-values may be tested. We mention that the foregoing method was suggested by Vich (1964), except that to speed up the calculation of f_l for higher l-values he transformed f_l, p_l, and q_l to functions of z by means of the transformation

$$Z\{f\} = \sum_{l=0}^{\infty} f_l z^{-l} = G(z), \text{ say.}$$

He was able to express $G(z)$ in terms of $Z\{p\}$ and $Z\{q\}$ and some boundary values of f, p, and q, the result being a quotient of two polynomials of degree $n + 2$ in z. The final step then consists of the application of the well-known method of synthetic division of two polynomials to obtain the successive f_l values (Lanczos, 1956).

Illustrations.   Two of the four methods of inversion described in this appendix assume a polynomial representation of $f(-a \ln x)$, $f(-a \ln \overline{1-x})$, or $f(-a \ln \overline{1-x^2})$, a third assumes $\phi(s)$ to be a polynomial in s^{-1}, and the last expresses $\phi(s)$ as the quotient of two polynomials. Since $\phi(s)$ depends on the whole range of values of $f(\cdot)$ and *vice versa*, it is intuitively obvious that all of these polynomials are likely to be of a fairly high order and that a good number of values of $\phi(s)$ will be needed for an adequate inversion.

We will illustrate this comment by attempting to invert the Laplace transform

$$\phi(s) = \exp\left(\frac{1}{\sqrt{1 + 2s}} - 1\right) \qquad (B.32)$$

for a single value of z, namely unity. This L.T. is actually that of $F'(x, 1)$ given by (2.33) with $\lambda = 1$ and

$$P'(y) = \frac{1}{\sqrt{2\pi}} e^{-y/2} y^{-\frac{1}{2}},$$

namely, chi-square with one degree of freedom. The mean of X is 1 and the variance 3. Since $dP^n*(x)/dx$ is chi-square with n degrees of freedom, an explicit series form can be derived for $F'(x, 1)$ from (2.33), and numerical calculations show that $F'(x, 1)\big|_{x=1} = 0.16223$ to five decimal places.

We first use Salzer's method which results in the inversion formula (B.18). It is based on the assumption that $\phi(s)$ is a polynomial in s^{-1} without constant term, but although this implies that $\phi(0) \to \infty$ it is not a significant feature of the polynomial fitting, since unity is the smallest value of s used. On the other hand $\phi(\infty) = 0$ for such a polynomial, and this ordinate is actually used in the Lagrange interpolation on which the method is based.

In the case of $\phi(s)$ given by (B.32) the ordinate at infinity is e^{-1}. It is thus more satisfactory to use (B.18) to invert

$$\phi_1(s) = e^{-s} \phi(s) = \exp\left(\frac{1}{\sqrt{1 + 2s}} - \overline{1 + s}\right),$$

noting that this implies that

$$f_1(z) = f(z - 1), \qquad z \geq 1.$$

The calculation of $f_1(2) = f(1)$ is given in Table B.1, the values of $A_j^{(10)}$ being taken from Salzer's (1958) tables. It will be seen that the use of eight-figure factors could produce an answer correct only to four decimal places but the result is, in fact, hopelessly incorrect. Clearly $\phi(s)$ does not lend itself to representation by a polynomial of the tenth degree in s^{-1}, but this is not surprising since, when $s > \frac{1}{2}$, it is actually an infinite series in powers of $s^{-\frac{1}{2}}$ with a constant term.

As a further example we choose the series (B.11), using the table of $S_n(\cdot)$ cited in Hochstrasser (1964). We have $U_n^*(x) = S_n(4x - 2)$, and since a may be chosen arbitrarily we simplify our reference to the tables by putting $e^{-1/a} = 0.625$ so that $a^{-1} = 0.4700036292$. The Table B.2 shows the values of $\phi[(k + 1)/a]$, c_k calculated from (B.12), using the coefficients of Lanczos' Table IX, and $U_k^*(0.625)$. The resulting values of $f(1)$ from (B.11) are shown in the last column, where $N = k$. It is clear that $f(-a \ln x)$ cannot be represented as a polynomial in x of any degree up to the eleventh.

Table B.1

j	$A_j^{(10)}(2)$	$(1 + 2j)^{-1}$	$\begin{array}{c} j+1 \\ -(1+2j)^{-1/2} \end{array}$	$\begin{array}{c} \exp\{-(4)\} \\ = \phi_1(j) \end{array}$	$A_j^{(10)}(2)\,\phi_1(j)$
(1)	(2)	(3)	(4)	(5)	(6)
1	$-0.0^334752841$	$0.\dot{3}$	1.42264973	0.24107439	-0.00008
2	3.0694684	0.2	2.55278640	0.077864402	0.23900
3	-567.29962	0.14285714	3.62203553	0.026728215	-15.16291
4	12,574.744	$0.\dot{1}$	4.6	$0.0^294035625$	118.24739
5	$-5,565.2867$	$0.\dot{0}\dot{9}$	5.69848866	$0.0^233510262$	-18.64942
6	$-476,224.46$	0.076923077	6.72264990	$0.0^212033452$	-573.06242
7	2,158,627.4	$0.0\dot{6}$	7.74180111	$0.0^343428867$	937.46742
8	$-3,917,642.7$	0.058823529	8.75746438	$0.0^315728291$	-616.17824
9	3,242,505.2	0.052631579	9.77058427	$0.0^457106976$	185.16967
10	$-1,014,196.8$	0.047619048	10.78178211	$0.0^420774545$	-21.06948
					$f(1) = f_1(2) \sim -2.99907$

In the particular case of (B.32) which is the L.T. of a density function, we may utilize the saddlepoint approximation described in Chapter 2. In order to apply formula (2.47), we first use (2.46) in the following manner:

$$\ln \phi(s) = K(u) = \frac{1}{\sqrt{1 - 2u}} - 1,$$

$$K'(u) = (1 - 2u)^{-3/2} = x,$$

$$K''(u) = 3(1 - 2u)^{-5/2},$$

$$K'''(u) = 15(1 - 2u)^{-7/2},$$

$$K^{iv}(u) = 105(1 - 2u)^{-9/2}.$$

When $x = 1$, $u = 0$, and the foregoing relations lead to

$$f(1) \simeq \frac{e^0}{\sqrt{6\pi}}\left[1 + \frac{1}{8} \cdot \frac{105}{3^2} - \frac{5}{24} \cdot \left(\frac{15}{3^{3/2}}\right)^2\right] = 0.16635,$$

Table B.2

k	$\phi\left(\dfrac{k+1}{a}\right)$	c_k	$U_k^*(0.625)$	$f(1)$
0	0.75424112	0.35449606	1.	0.43703
1	0.66315546	0.53774976	0.5	0.76850
2	0.61363380	0.69108243	-0.75	0.12952
3	0.58178626	0.86232965	-0.875	-0.80068
4	0.55925629	1.03155020	0.3125	-0.40328
5	0.54230538	1.20348313	1.03125	1.12675
6	0.52899047	1.37483002	0.203125	1.47103
7	0.51819336	1.54734234	-0.9296875	-0.30242
8	0.50922105	1.71916070	-0.66796875	-1.71811
9	0.50161908	1.88705889	0.59570312	-0.33228
10	0.49507594	2.11082638	0.96582031	2.18102
11	0.48937024	1.94356074	-0.11279297	1.91077

which is four units in error in the third decimal place. We note that

$$K^{(j)}(0)/K''(0) = (2j - 1)/3$$

so that the conditions for successful application of this method are not satisfied.

Finally, let us see the result of fitting a member of the Pearson family of curves by means of the first four moments. Since the rth cumulant of the chi-square distribution with 1 d.f. is $2^{r-1}(r - 1)!$, we have

$$p_1 = 1, \qquad p_2 = 2 + 1^2 = 3, \qquad p_3 = 8 + 3 \times 2 \times 1 + 1^3 = 15$$

and

$$p_4 = 48 + 4 \times 8 \times 1 + 3 \times 2^2 + 6 \times 2 \times 1^2 + 1^4 = 105.$$

The four moments of $F(\cdot, 1)$ are thus given by

$$\kappa_1 = p_1 = 1, \qquad \mu_2 = \kappa_2 = p_2 = 3, \qquad \mu_3 = \kappa_3 = p_3 = 15$$

and

$$\mu_4 = \kappa_4 + 3\kappa_2^2 = p_4 + 3p_2^2 = 105 + 3 \times 9 = 132,$$

which lead to

$$\beta_1 = \frac{\mu_3^2}{\mu_2^3} = \frac{225}{27} = 8.\dot{3} \quad \text{and} \quad \beta_2 = \frac{\mu_4}{\mu_2^2} = \frac{132}{9} = 14.\dot{6}.$$

Reference to the (β_1, β_2) chart in Pearson (1931) shows that a Type I curve is applicable, and if we use Elderton's (1938, cited in Chapter 2) form with origin at the mean, namely,

$$y = y_e \left(1 + \frac{x}{A_1}\right)^{m_1} \left(1 - \frac{x}{A_2}\right)^{m_2}, \tag{B.33}$$

we have (in his notation)

$$r = 6 \frac{\beta_2 - \beta_1 - 1}{6 + 3\beta_1 - 2\beta_2} = 6 \frac{5.3}{1.6} = 19.2$$

$$A_1 + A_2 = \tfrac{1}{2}\sqrt{\mu_2} \left[\beta_1(r + 2)^2 + 16(r + 1)\right]^{1/2}$$

$$= 0.86602540(3745.3 + 323.2)^{1/2}$$

$$= 0.86602540 \times 63.785056 = 55.239479$$

$$\left.\begin{matrix} m_2 \\ m_1 \end{matrix}\right\} = \tfrac{1}{2}\left\{r - 2 \pm r(r + 2)\left[\frac{\beta_1}{\beta_1(r + 2)^2 + 16(r + 1)}\right]^{1/2}\right.$$

$$= \tfrac{1}{2}\left(17.2 \pm \frac{407.04 \times 2.8867513}{63.785056}\right) = \frac{1}{2}\left(\begin{matrix} 35.621607 \\ -1.221607 \end{matrix}\right)$$

$$m_1 = -0.610804, \qquad m_2 = 17.810804,$$

and finally, since we do not need to separate A_1 and A_2,

$$
\begin{aligned}
y_e &= \frac{1}{A_1 + A_2} \cdot \frac{(m_1 + 1)^{m_1}(m_2 + 1)^{m_2}}{(m_1 + m_2 + 2)^{m_1+m_2}} \cdot \frac{\Gamma(m_1 + m_2 + 2)}{\Gamma(m_1 + 1)\,\Gamma(m_2 + 1)} \\
&= 0.018102995 \times \frac{(0.389196)^{-0.610804}(18.810804)^{17.810804}}{(19.2)^{17.2}} \\
&\quad\quad \cdot \frac{\Gamma(19.2)}{\Gamma(0.389196)\,\Gamma(18.810804)} \\
&= 0.018102995 \times \text{antilog}\,(1.0108118) = 0.185593.
\end{aligned}
$$

Putting $x = 0$ in (B.32), we have

$$
f(1) \simeq y_c = 0.18559,
$$

which is a long way from the true value 0.16223.

INVERSION OF A CHARACTERISTIC FUNCTION

We conclude by describing Bohman's (1963) ingenious method for inverting the characteristic function of a probability density. It requires the use of an electronic computer and was employed very successfully by Bohman and Esscher (1963; cited in Chapter 2) as a standard by which to judge various approximate formulas for $F(x, t)$ and ρ_x.

If $\psi(u)$ is the characteristic function of the probability density $f(\cdot)$, then (u real)

$$
\psi(u) = \int_{-\infty}^{\infty} e^{iux} f(x)\, dx = \int_{-\infty}^{\infty} e^{iux}\, dF(x). \tag{B.34}
$$

On comparing this with the L.T. of $f(x)$, supposing that $f(x) = 0$, $x < 0$, we see that

$$
\psi(u) = \phi(-iu).
$$

Bohman's argument proceeds in two stages: in the first he shows how to bracket a d.f. by means of two functions $F_1(\cdot)$ and $F_2(\cdot)$, and in the second he chooses a bounded symmetrical function $C(\cdot)$ which facilitates the inversion of the characteristiclike functions of $F_1(\cdot)$ and $F_2(\cdot)$. Consider the convolution ($j = 1, 2$)

$$
F_j(x) = \int_{-\infty}^{\infty} F(x - y)\, h_j(y)\, dy \tag{B.35}
$$

and define

$$
H_j(x) = \begin{cases} \displaystyle\int_x^{\infty} h_j(y)\, dy, & x > 0, \\[2mm] \displaystyle -\int_{-\infty}^{x} h_j(y)\, dy, & x < 0, \end{cases} \tag{B.36}
$$

so that

$$
H_j(0+) - H_j(0-) = \int_{-\infty}^{\infty} h_j(y)\, dy \equiv 1.
$$

Integrating (B.35) by "parts,"

$$F_j(x) = -F(x-y) H_j(y) \Big|_{-\infty}^{0-} - F(x-y) H_j(y) \Big|_{0+}^{\infty} + \int_{-\infty}^{\infty} H_j(y)\, d_y F(x-y)$$

$$= F(x) - \int_{-\infty}^{\infty} H_j(x-y)\, dF(y).$$

Thus $F_j(x) \gtrless F(x)$ according as $H_j(y) \gtrless 0$ for all y. We now find $H_1(\cdot)$ and $H_2(\cdot)$ such that

$$F_1(x) < F(x) < F_2(x). \qquad (B.37)$$

Consider the characteristiclike functions

$$C_j(u) = C(u) - k_j i\, C'(u), \qquad k_j = \begin{cases} 0.42, & j = 1, \\ -0.42, & j = 2, \end{cases} \qquad (B.38)$$

where

$$C(u) = \begin{cases} (1 - |u|)\cos \pi u + \dfrac{1}{\pi}\sin |\pi u|, & |u| \le 1, \\ 0, & |u| > 1, \end{cases} \qquad (B.39)$$

and thus

$$C'(u) = \begin{cases} -\pi(1 - |u|)\sin \pi u, & |u| < 1, \\ 0, & |u| > 1. \end{cases} \qquad (B.40)$$

It may be verified that the function

$$h_j(y) = 2\pi \frac{1 + \cos y}{(y^2 - \pi^2)^2} (1 - k_j y)$$

satisfies our requirements on $H_j(x)$ and that

$$C_j(u) = \int_{-\infty}^{\infty} e^{iuy} h_j(y)\, dy.$$

We note in particular that

$$\int_0^{\infty} 4\pi y \frac{1 + \cos y}{(y^2 - \pi^2)^2}\, dy = 2\int_0^{\pi} \frac{\sin y}{y}\, dy - \frac{4}{\pi} < 0.42;$$

$F_j(x)$ is then a function obtained by inverting the characteristiclike function of the convolution (B.35), namely, by inverting the product $\psi(u)\, C_j(u)$.

It is convenient to write $\psi(u)$ in terms of the two real functions $A(u)$ and $B(u)$ such that

$$\psi(u) = A(u) e^{iB(u)},$$

where $A(-u) = A(u)$ and $B(-u) = -B(u)$. Then the standard Fourier inversion formula (Cramér, 1946, cited in Chapter 2) provides

$$f_j(x) = \frac{1}{2\pi} \int_{-T}^{T} A(u) e^{i(B(u)-xu)} C_j\left(\frac{u}{T}\right) du$$

$$= \frac{1}{\pi} \int_0^T C\left(\frac{u}{T}\right) A(u) \cos\left(xu - B(u)\right) du$$

$$- \frac{k_j}{\pi} \int_0^T C'\left(\frac{u}{T}\right) A(u) \sin\left(xu - B(u)\right) du, \qquad \text{(B.41)}$$

and thus, on integrating over x,

$$F_j(x) = \frac{1}{2} + \frac{1}{\pi} \int_0^T C\left(\frac{u}{T}\right) A(u) \frac{\sin\left(xu - B(u)\right)}{u} du$$

$$- \frac{k_j}{\pi} \int_0^T C'\left(\frac{u}{T}\right) A(u) \frac{\cos\left(xu - B(u)\right)}{u} du, \quad \text{(B.42)}$$

where T is chosen suitably. We notice that the function corresponding to $C_j\left(\frac{u}{T}\right)$ is $\varepsilon^{-1} h_j\left(\frac{y}{\varepsilon}\right)$, $\varepsilon = T^{-1}$, and that (B.35) becomes

$$F_j(x) = \int_{-\infty}^{\infty} F(x - y) \varepsilon^{-1} h_j\left(\frac{y}{\varepsilon}\right) dy = \int_{-\infty}^{\infty} F(x - \varepsilon y)\, h_j(y)\, dy$$

$$\sim F(x) - \varepsilon F'(x) \int_{-\infty}^{\infty} y\, h_j(y)\, dy.$$

Hence $F_1(x)$ and $F_2(x)$ both converge to $F(x)$ as $T \to \infty$.

The solution of the problem of inverting the characteristic function $\psi(u)$ defined by (B.34) is thus achieved through the inequalities (B.37), where $F_1(\cdot)$ and $F_2(\cdot)$ are provided by (B.42). The computer is, of course, required for the evaluation of (B.42) for large T, using an appropriate quadrature formula. Bohman and Esscher (*loc. cit.*), for example, chose $T = 90\pi$ and used the trapezoidal rule with an interval of $\pi/12$, that is, with 1081 ordinates. It may be added that they worked in standard measure to avoid large values of x.

As an illustration, we choose the characteristic function corresponding to (B.32), namely,

$$\psi(u) = \exp\left[(1 - 2iu)^{-\frac{1}{2}} - 1\right],$$

so that

$$a(u) + i B(u) = (1 - 2iu)^{-\frac{1}{2}} - 1, \qquad \text{(B.43)}$$

where

$$A(u) = e^{a(u)}.$$

On comparing the real and imaginary parts of relation (B.43)

$$(1 + a)^2 - B^2 + 4u(1 + a)B = 1$$

and

$$u[(1 + a)^2 - B^2] = (1 + a)B.$$

Hence

$$(1 + a)^2 - B^2 = \frac{1}{1 + 4u^2}$$

and

$$(1 + a)B = \frac{u}{1 + 4u^2}.$$

The resulting quadratic in $(1 + a)^2$ provides a real value for a such that

$$(1 + a)^2 = \frac{1 + \sqrt{1 + 4u^2}}{2(1 + 4u^2)}$$

and this conforms with the requirement that $A(-u) = A(u)$. It follows that

$$\ln A(u) = a(u) = \left[\frac{1 + \sqrt{1 + 4u^2}}{2(1 + 4u^2)}\right]^{\frac{1}{2}} - 1 \qquad (B.44)$$

and

$$B(u) = \frac{u}{1 + 4u^2}[1 + a(u)]^{-1}.$$

Using the trapezoidal rule with 1081 ordinates, $x = 1$ and $T = 90\pi$, relations (B.42),* with $C(\cdot)$, $C'(\cdot)$, $A(\cdot)$, and $B(\cdot)$ given by (B.39), (B.40), and (B.44), provided the results 0.74437 and 0.74914, respectively, in eight seconds of IBM 7094/7040 Direct Couple computer execution time (compilation required less than half a minute). We may thus write

$$F(1, 1) = 0.7468 \pm 0.0024.$$

This is the same kind of "maximum" deviation that Bohman and Esscher (*loc. cit.*) obtained in the neighborhood of the means of their distributions. That the indicated error is greatly overstated may be seen from the mean of the two results obtained by using relations (B.41), namely

$$f(1, 1) \simeq \tfrac{1}{2}(0.16053 + 0.16396) = 0.16224,$$

which is only a unit in error in the fifth decimal place.

A practical problem in the utilization of Bohman's method is the size to choose for T and the number of ordinates to be used in the trapezoidal rule

---

* Note that the ordinates at $u = 0$ in the two integrals appearing therein are, by a limiting process, x and π/T, respectively.

to achieve a given degree of accuracy in the evaluation of the resulting integrals. Bohman and Esscher (*loc. cit*) show how to evaluate the error introduced by quadrature, and Romberg's method of improvement of trapezoidal approximations effected at intervals subjected to successive halving (Wilf, 1967) implements these ideas. There is, however, no guarantee of success. Using $2^{10} = 1024$ subintervals instead of 1080 in the foregoing illustration the trapezoidal rule produced $F(1, 1) = .7490 \pm .0024$ while 2048 subintervals resulted in $F(1, 1) = .7271 \pm .0024$. On the other hand repeated Simpson with 2048 subintervals provided $F(1, 1) = .7197 \pm .0024$ and the last of the Romberg extrapolates gave $F(1, 1) = .7184 \pm .0024$.

REFERENCES

Bateman, H. (1936). "Two systems of polynomials for the solution of Laplace's integral equation." *Duke Math. J.*, **2**, 569–577.

Bellman, R., R. E. Kalaba, and J. A. Lockett (1966). *Numerical Inversion of the Laplace Transform*. Elsevier, New York.

Berger, B. (1966). "Inversion of the *n*-dimensional Laplace transform." *Math. Comp.*, **20**, 418–421.

Bohman, H. (1963). "To compute the distribution function when the characteristic function is known." *Skand Aktuar. Tidskr*, **46**, 41–46.

Bowers, N. L., Jr. (1966). "Expansion of probability density functions as a sum of gamma densities with applications in risk theory." *Trans. Soc. Actu.*, **18**, 125–147.

Davis, P. J., and I. Polonsky (1964). "Numerical interpolation, differentiation and integration." *Handbook of Mathematical Functions*, M. Abramowitz and I. A. Stegun (Eds.), National Bureau of Standards, Washington, D.C.

Erdélyi, A. (1943). "Inversion formulae for the Laplace transformation." *Phil. Mag.*, Ser. 7, **34**, 533–537.

Hochstrasser, U. W. (1964). "Orthogonal polynomials." *Handbook of Mathematical Functions*, M. Abramowitz and I. A. Stegun (Eds.), National Bureau of Standards, Washington, D.C.

Khamis, S. H. (1960). "Incomplete gamma functions expansions of statistical distribution functions." *Bull. Inst. Internat. Statist.*, **37**, 385–396.

Lanczos, C. (1956). *Applied Analysis*. Prentice-Hall, Englewood Cliffs, N.J.

Miller, M. K., and W. T. Guy (1966). "Numerical inversion of the Laplace transform by use of Jacobi polynomials." *SIAM J. Numer. Anal.*, **3**, 624–635.

Papoulis, A. (1956). "A new method of inversion of the Laplace transform." *Quart. Appl. Math.*, **14**, 405–414.

Pearson, K. (Ed.) (1931). *Tables for Statisticians and Biometricians*, Part II. Cambridge University Press, Cambridge.

Salzer, H. (1958). "Tables for the numerical calculation of inverse Laplace transforms." *J. Math. Phys.*, **37**, 89–109.

Stroud, A. H., and D. Secrist (1966). *Gaussian Quadrature Formulas*. Prentice-Hall, Englewood Cliffs, N.J.

Vich, R. (1964). *Z-Transformation: Theorie und Anwendung*. VEB, Berlin.

Wilf, H. S. (1967). "Advances in numerical quadrature." *Mathematical Methods for Digital Computers*, Vol. II, A. Ralston and H. S. Wilf (Eds.). Wiley, New York, pp. 133–144.

Glossary of Consistently Used Notation

Page of First Appearance	Symbol	Meaning
4	$\mathcal{P}\{\cdot\}$	The probability of the event enclosed in the braces.
4	X	The random variable (r.v.) representing the aggregate claim outgo of a risk business or, sometimes, of a single contractholder.
4	$F(\cdot)$	The distribution function (d.f.) of X.
5	R	The random variable representing the loss ratio X/S, where S is the aggregate sum insured of a risk business with claim outgo X.
5	$\Gamma(\cdot)$	Gamma function.
8	κ_j	The jth cumulant of $F(\cdot)$; κ_1 is the mean value of X.
9	$\Phi(\cdot)$	The d.f. of a Normal random variable with zero mean and unit variance.
10	$\mathcal{E}(\cdot)$	The expected value of a random variable.
10	$\mathcal{V}(\cdot)$	The variance of a random variable.
12	$p_n(t)$	The probability of n claims occurring within a time interval of length t.
12	Y	The positive random variable representing the size of an individual claim independent of the random variable representing the number of claims. Although Y may depend on the epoch at which the individual claim occurs, it is generally assumed to be time invariant.
12	$P(\cdot)$	The d.f. of Y; $P(0) = 0$.
12	$P^{n*}(\cdot)$	The d.f. of the sum of n independent r.v.'s each distributed as Y; $P^{n*}(0) = 0$.
12	$P^{0*}(\cdot)$	Defined on p. 12.

201

Page of First Appearance	Symbol	Meaning
12	$F(x, t)$	The d.f. of the aggregate X in an interval of length t.
13	$M_Y(\theta)$	Moment generating function (mgf) of Y.
14	$\lambda(\tau)$	The instantaneous rate (force) of claim at time τ; also called the hazard at time τ. If this function is a constant, τ is suppressed.
17	Λ	The time-invariant r.v. of the accident hazard.
17	$U(\cdot)$	The d.f. of Λ, called the mixing distribution of accident hazards. [Not to be confused with the $U(\cdot)$ fleetingly used in Chapter 4 in the solution of the general renewal equation.]
18	$k^{(j)}$	The factorial $k(k - 1)(k - 2) \cdots (k - j + 1), j \leq k$.
19	$\mathfrak{R}(s)$	The real part of the complex number s.
19	$\mathfrak{L}\{\ \}$	The Laplace transform (L.T.) of the function inside the braces.
29	$\mu(y)$	$P'(y)/[1 - P(y)]$.
30	Pareto law	$P(y) = 1 - (y/y_0)^{-\alpha}, y_0 < y < \infty$.
30	p_j	The jth moment of Y about zero, $j = 0, 1, 2, \ldots$.
49	$\kappa_1, \hat{\kappa}_1$	The aggregate net premium of a risk business during a unit interval of time and its estimate.
50	ρ_c	The net premium for aggregate claims in excess of "priority" c.
50	κ_{1c}	The mean of the aggregate claims in excess of "priority" c.
56	$\pi_1 = \kappa_1 + \eta_1$ $= \kappa_1(1 + \eta)$	The risk-loaded premium for a given time interval which may tend to zero; η_1 is the aggregate risk loading and η, the rate of loading.
65	$p_{l\|n}(s \mid t)$	The probability of l claims occurring during a period of length s following a period of length t during which there were n claims.
78	p_{ij}	The probability of a transition from state i to state j of a Markov chain.

Page of First Appearance	Symbol	Meaning
78	$p_{ij}^{(t)}$	The probability of a transition from state i to state j in t unit time intervals.
83	Z	The "weight" or "credibility index" of one or more contracts (Chapter 3 only).
91	N_τ	The number of contractholders at time t.
92	T	The random variable representing the time to ruin in a suitably chosen time scale. One such scale is the expected interval between successive claims.
92	$Y(t)$	The random variable representing the aggregate claims made in a period $(0, t)$.
92	$R(t)$	The risk reserve at time t. Various notations (e.g., w, x, z) are used for $R(0)$.
93	$\psi(w)$ $= 1 - \phi(w)$	The probability of ultimate ruin, given that $w \geq 0$ is the current risk-reserve.
94	$\Phi(s)$	The Laplace transform of $\phi(\cdot)$.
96, 104	$v(x, t)$ $= 1 - u(x, t)$	The probability of ruin within a period $(0, t)$, given that $x \geq 0$ is the initial risk-reserve.
110	$\pi(s)$	The Laplace-Stieltjes transform of $P(\cdot)$.
110	$h(\cdot) =$ $(1 - P(\cdot))\pi_1^{-1}$	
110	$\chi(s)$	The Laplace transform of $h(\cdot)$.
110	$\mu(r, s)$	The bivariable Laplace transform of $u(x, t)$.
111	$\mu(s)$	The Laplace transform of $u(0, t)$.
112	$a_k(x)$	kth moment of time to ruin.
113, 130	$R_0 = -\kappa$	A quantity defined by $\int_0^\infty e^{\kappa x} h(x)\, dx = 1$.
116	$B(\cdot)$	The d.f. of immediate life annuity values; $B(0) = 0$.
116	$\pi_2 = p_1 - \eta_1$	The mean annuity outgo per unit of time.
122	Z	The ceiling sometimes imposed on the risk-reserve (Chapters 5 and 6).
123	$Z(\cdot)$	The unknown function in the general renewal equation.
137	M	The maximum sum-at-risk retained on an individual contract.
159	$u(x)$	The utility of a sum of money x.
159	$U(R, F)$	The utility index of a risk business with capital R and claim d.f. $F(\cdot)$.

Page of First Appearance	Symbol	Meaning
164	$D(R)$	Expected number of years to the ruin of a risk business with capital R and claim d.f. $F(\cdot)$.
165	$V(R)$	Value of future dividends when capital is R and claim d.f. is $F(\cdot)$.
174	$N(t)$	The random variable representing the number of (renewal) events occurring in an interval of length t following an arbitrary epoch.

Author Index

Subject Index